SpringerBriefs in Electrical and Computer Engineering

Series Editors

Woon-Seng Gan, School of Electrical and Electronic Engineering, Nanyang Technological University, Singapore, Singapore

C.-C. Jay Kuo, University of Southern California, Los Angeles, CA, USA

Thomas Fang Zheng, Research Institute of Information Technology, Tsinghua University, Beijing, China

Mauro Barni, Department of Information Engineering and Mathematics, University of Siena, Siena, Italy

SpringerBriefs present concise summaries of cutting-edge research and practical applications across a wide spectrum of fields. Featuring compact volumes of 50 to 125 pages, the series covers a range of content from professional to academic. Typical topics might include: timely report of state-of-the art analytical techniques, a bridge between new research results, as published in journal articles, and a contextual literature review, a snapshot of a hot or emerging topic, an in-depth case study or clinical example and a presentation of core concepts that students must understand in order to make independent contributions.

More information about this series at http://www.springer.com/series/10059

Yannick Deville • Leonardo Tomazeli Duarte
Shahram Hosseini

Nonlinear Blind Source Separation and Blind Mixture Identification

Methods for bilinear, linear-quadratic and polynomial mixtures

 Springer

Yannick Deville 🅞
Paul Sabatier University
Toulouse, France

Leonardo Tomazeli Duarte
State University of Campinas
Campinas, Brazil

Shahram Hosseini
Paul Sabatier University
Toulouse, France

ISSN 2191-8112   ISSN 2191-8120   (electronic)
SpringerBriefs in Electrical and Computer Engineering
ISBN 978-3-030-64976-0   ISBN 978-3-030-64977-7   (eBook)
https://doi.org/10.1007/978-3-030-64977-7

This Springer imprint is published by the registered company Springer Nature Switzerland AG
The registered company address is: Gewerbestrasse 11, 6330 Cham, Switzerland

Preface

Blind source separation methods aim at estimating a set of source signals, which have unknown values but some known properties, from a set of observed signals that are mixtures of these source signals, with partly unknown mixing transforms. Most of the methods proposed so far are restricted to the simplest class of mixtures, namely linear ones (instantaneous or not). However, various more advanced studies dealing with nonlinear mixtures have also been reported. A large part of these investigations concern (1) linear-quadratic mixtures, including their bilinear and purely quadratic restricted forms, and (2) to a much lower extent, polynomial mixtures. These mixtures form natural extensions of linear ones, by also including second-order and possibly higher-order terms. They may thus first be seen as generic models, to be used as approximations (truncated polynomial series) of various, possibly unknown, models faced in practical applications (e.g., arrays of ion-selective electrodes). Moreover, linear-quadratic mixing has been shown to actually occur in different applications, namely unsupervised unmixing of remote sensing hyperspectral (and multispectral) images, processing of scanned images involving the show-through effect, and analysis of gas sensor array data. Based on the increasing attractiveness of these topics, this book provides a survey of these nonlinear mixing models, associated separating structures and methods in terms of blind source separation and blind mixture identification, and corresponding properties (invertibility, identifiability/separability). Independent component analysis and Bayesian methods, matrix factorization approaches with or without nonnegativity constraints, and sparse component analysis are thus especially considered.

Acknowledgments

The authors would like to thank all their co-workers on the topics addressed in this book, whose names mostly appear in the list of references at the end of the book.

Contents

Chapter 1
Introduction

Blind source separation (BSS) methods aim at estimating a set of source signals, which have unknown values but some known properties, from a set of observed signals that are mixtures of these source signals (see, e.g., [22, 24, 28, 31, 64, 92]). It has been shown that, if the mixing function applied to the source signals is completely unknown, the class of solutions to the BSS problem based on independent component analysis (ICA) [24, 94] leads to unacceptable so-called indeterminacies, i.e., unavoidable residual transforms in the estimated source signals, as compared with the actual source signals. Therefore, in most investigations the mixing function is requested to belong to a known class and only the values of its parameters are to be estimated. Many of these works are restricted to the simplest class of mixtures, namely linear ones (instantaneous or not). However, various more advanced studies dealing with *nonlinear* mixtures have also been reported. A large part of these investigations concern (1) linear-quadratic mixtures, including their bilinear and purely quadratic restricted forms, and (2) to a much lower extent, polynomial mixtures. These mixtures form natural extensions of linear ones, by also including second-order and possibly higher-order terms. They may thus first be seen as generic models, to be used as approximations (truncated polynomial series) of possibly unknown models faced in various fields: see, e.g., their application to arrays of ion-selective electrodes in [8, 9]. Moreover, linear-quadratic mixing has been shown to *actually* occur in different applications. It has thus mainly been used for unmixing remote sensing[1] data (see, e.g., [14, 45, 51, 63, 65, 66, 79–82, 91, 93]),

---

[1]Blind source separation is often called unsupervised (spectral) unmixing by the remote sensing community. In this unsupervised unmixing problem, a linear-quadratic mixing model is obtained when only considering direct and second-order reflections. As discussed, e.g., in [93, 96] and in Chap. 7 of the present book, a polynomial model is obtained when also considering higher-order reflections. However, it is often argued that these higher-order terms can be neglected.

Y. Deville et al., *Nonlinear Blind Source Separation and Blind Mixture Identification*, SpringerBriefs in Electrical and Computer Engineering, https://doi.org/10.1007/978-3-030-64977-7_1

processing scanned images involving the show-through effect [6, 7, 43, 47, 74, 84], and analyzing gas sensor array data [11, 76, 77].

In this book, we provide a survey of the above-defined nonlinear mixing models and associated BSS methods reported so far. To this end, we consider the standard procedure for developing BSS methods that was proposed, e.g., in [28, 31]. It consists in defining the following items for each version of the proposed methods:

1. The considered source signals, observations, and mixing model
2. The proposed separating system structure (this may require one to first analyze the invertibility of the considered mixing model)
3. The proposed separation principle (e.g., the independence principle, which consists in making the separating system outputs statistically independent; this principle is used in ICA)
4. The proposed separation criterion (e.g., minimization of a given cost function, such as output mutual information in ICA)
5. The proposed separation algorithm (e.g., gradient descent, for a given cost function)

This book is organized accordingly. In Chap. 2, we introduce the expressions and variants of the main mixing model considered in this book, namely the linear-quadratic model (including its bilinear version). The invertibility of that model is analyzed in Chap. 3, where we also describe the separating structures that have been proposed to handle signals mixed according to that model. The last three steps of the above-defined procedure concern the adaptation of the considered separating system. A separate chapter of this book is then essentially allocated to each reported separation principle and to the resulting class of linear-quadratic BSS criteria and algorithms: see Chaps. 4–6. The extension of all these topics to general polynomial mixtures is eventually briefly presented in Chap. 7, together with conclusions drawn from this survey.

Besides, as discussed, e.g., in [28, 29, 31], the blind mixture identification (BMI) problem is closely related to BSS. Comments about BMI are therefore also provided throughout this book.

Chapter 2
Expressions and Variants of the Linear-Quadratic Mixing Model

2.1 Scalar Form of Mixing Model

Unless otherwise stated, all signals considered in this book depend on a discrete variable n and are continuous-valued and real-valued. The simplest class of mixtures that may be defined for such a set of source signals consists of linear memoryless (or instantaneous) mixtures, which read[1]

$$x_i(n) = \sum_{j=1}^{M} a_{ij} s_j(n) \qquad \forall\, i \in \{1, \ldots, P\}, \tag{2.1}$$

where $x_i(n)$ are the values of the P observed mixed signals for the sample index n and $s_j(n)$ are the values of the M unknown source signals which yield these observations, whereas all a_{ij} are the unknown (linear) mixing coefficients which define the considered source-to-observation transform. Each value $x_i(n)$ of an observed signal for a given sample index n is thus a *linear combination* of the values $s_j(n)$ of the source signals for the *same* index n (i.e., no memory effect along the series of values indexed by n), hence the name of this mixing model.

In most of this book, we consider another class of mixtures, which is a superset of the above class and which reads

$$x_i(n) = \sum_{j=1}^{M} a_{ij} s_j(n) + \sum_{j=1}^{M} \sum_{k=j}^{M} b_{ijk} s_j(n) s_k(n) \qquad \forall\, i \in \{1, \ldots, P\}. \tag{2.2}$$

---

[1] In this book, we consider noiseless mixtures, unless otherwise stated.

Y. Deville et al., *Nonlinear Blind Source Separation and Blind Mixture Identification*, SpringerBriefs in Electrical and Computer Engineering, https://doi.org/10.1007/978-3-030-64977-7_2

It is thus also memoryless but, in addition to the linear terms $a_{ij}s_j(n)$ faced in the previous model, it contains the terms $b_{ijk}s_j(n)s_k(n)$. Each of these terms corresponding to $j = k$ is quadratic with respect to one source signal, so that this mixing model (2.2) is stated to be linear-quadratic. All b_{ijk} are supposedly unknown and are called the "quadratic mixing coefficients." Unless otherwise stated, all coefficients a_{ij} and b_{ijk} are continuous-valued and real-valued in this book. In its most general form (2.2), this model contains two types of second-order terms:

1. The second-order auto-terms are the terms which involve a single source signal. They correspond to $j = k$ and they read $b_{ijj}[s_j(n)]^2$.
2. The second-order cross-terms are the terms which involve two different source signals. They correspond to $j \neq k$ (and hence $j < k$) and they read as in (2.2), i.e., $b_{ijk}s_j(n)s_k(n)$, keeping in mind that $s_j(n)$ and $s_k(n)$ are then not the same signal.

A subclass of the linear-quadratic mixing model therefore consists of mixtures which contain no second-order auto-terms (i.e., $b_{ijk} = 0$ when $j = k$) and which then read[2,3]

$$x_i(n) = \sum_{j=1}^{M} a_{ij}s_j(n) + \sum_{j=1}^{M-1}\sum_{k=j+1}^{M} b_{ijk}s_j(n)s_k(n) \quad \forall i \in \{1, \ldots, P\}. \quad (2.3)$$

It should be noted that, although each cross-term $b_{ijk}s_j(n)s_k(n)$ of (2.3) is of second order with respect to the overall pair of signals $\{s_j(n), s_k(n)\}$, each such term and therefore the overall model (2.3) remains linear with respect to each source signal considered separately (as opposed to the general model (2.2) also involving quadratic auto-terms). This model (2.3) is therefore often referred to as the bilinear (again memoryless) mixing model. Its simplest version is obtained for $P = 2$ mixtures of $M = 2$ source signals and reads

$$x_1(n) = a_{11}s_1(n) + a_{12}s_2(n) + b_1 s_1(n)s_2(n) \quad (2.4)$$

$$x_2(n) = a_{21}s_1(n) + a_{22}s_2(n) + b_2 s_1(n)s_2(n) \quad (2.5)$$

when using the simplified notations b_1 and b_2 for the quadratic mixing coefficients denoted as b_{ijk} in (2.3).

A subclass of the bilinear model (2.3), often referred to as the "Fan Model," was also introduced in [46] for some remote sensing applications involving forest. It is obtained by constraining each quadratic coefficient b_{ijk} of (2.3), associated with the

---

[2]The sums $\sum_{j=1}^{M}\sum_{k=j}^{M}$ and $\sum_{j=1}^{M-1}\sum_{k=j+1}^{M}$ in (2.2) and (2.3) may be expressed in more compact form as $\sum_{1 \leq j \leq k \leq M}$ and $\sum_{1 \leq j < k \leq M}$, respectively.

[3]Additional constant terms are considered in [6, 7] for the bilinear model. Besides, the linear-quadratic model in [76] contains constant terms equal to 1, which can be subtracted from the observations, whereas the linear-quadratic model in [77] contains arbitrary constant terms.

pair of source signals $\{s_j(n), s_k(n)\}$, to be equal to the product of the two linear coefficients a_{ij} and a_{ik}, respectively, associated with these two source signals. The mixing model thus also becomes bilinear with respect to the mixing coefficients, and is referred to as the bilinear-bilinear model in [45]. An extension of that model was proposed in [54] and then used, e.g., in [91].

Other subclasses of the general linear-quadratic model (2.2) or of its other forms defined above are also considered in some application fields, because in these fields the source values and/or mixing coefficients are guaranteed to meet some properties. This especially includes applications where the source values and/or mixing coefficients are always nonnegative, and those where, separately for each observed signal x_i, the sum of the values of all linear coefficients a_{ij} is equal to one (see (5.4)).

Finally, another subclass within the linear-quadratic mixing model consists of quadratic mixtures, which are defined by (2.2) but which contain no linear terms $a_{ij}s_j(n)$. Their simplest version is again obtained for $P = 2$ mixtures of $M = 2$ source signals and reads

$$x_1(n) = b_{111}[s_1(n)]^2 + b_{112}s_1(n)s_2(n) + b_{122}[s_2(n)]^2 \qquad (2.6)$$

$$x_2(n) = b_{211}[s_1(n)]^2 + b_{212}s_1(n)s_2(n) + b_{222}[s_2(n)]^2. \qquad (2.7)$$

The general linear-quadratic model and its bilinear subclass have been considered in various application fields in the literature, as mentioned in Chap. 1. The quadratic subclass has only been studied by a few authors, especially by Chaouchi et al. [18–21].[4] These quadratic mixtures will, e.g., possibly become of importance in applications where sensors provide observed signals $x_i(n)$ which are sensitive to (i.e., linear combinations of) the individual powers $[s_j(n)]^2$ of the source signals and to their "cross-powers" $s_j(n)s_k(n)$, but not directly to their magnitudes $s_j(n)$, as in (2.6)–(2.7).

2.2 Matrix-Vector Forms of Mixing Model

For linear mixtures, initially defined by (2.1), a matrix form of that model (2.1) is usually derived by introducing the source and observation vectors

$$s(n) = [s_1(n), \ldots, s_M(n)]^T \qquad (2.8)$$

$$x(n) = [x_1(n), \ldots, x_P(n)]^T, \qquad (2.9)$$

---

[4]In these papers, the mixing model is expressed in a slightly different way than in (2.6)–(2.7). This is equivalent to restricting oneself to the case when the coefficients $b_{111}, b_{122}, b_{211}$, and b_{222} of the auto-terms of (2.6)–(2.7) are nonnegative, as discussed further in this book: see (4.9)–(4.10).

where ^{T} stands for transpose. Equation (2.1) then yields

$$x(n) = As(n), \qquad (2.10)$$

where the $P \times M$ matrix A consists of the mixing coefficients a_{ij}.

That approach also directly applies to the linear part of the linear-quadratic model (2.2). Its quadratic part may be independently transformed by using the same approach. To this end, one builds:

- a column vector $p(n)$, composed of all source products $s_j(n)s_k(n)$ of (2.2), i.e., with $1 \le j \le k \le M$ arranged in a fixed, arbitrarily selected, order (see, e.g., [82] for the natural order),
- a matrix B, composed of all entries b_{ijk} arranged so that i is the row index of B and the columns of B are indexed by (j, k) and arranged in the same order as the source products $s_j(n)s_k(n)$ in $p(n)$.

The linear-quadratic mixing model (2.2) may then be expressed in matrix form as

$$x(n) = As(n) + Bp(n). \qquad (2.11)$$

An even more compact form of this model may be derived by stacking row-wise the vectors $s(n)$ and $p(n)$ of sources and source products in an extended vector

$$\tilde{s}(n) = \begin{bmatrix} s(n) \\ p(n) \end{bmatrix}, \qquad (2.12)$$

whereas the corresponding matrices A and B are stacked column-wise in an extended matrix

$$\tilde{A} = [A \quad B]. \qquad (2.13)$$

The linear-quadratic mixing model (2.11) then yields

$$x(n) = \tilde{A}\tilde{s}(n). \qquad (2.14)$$

Comparing this model (2.14) to (2.10) shows that the vector $x(n)$, initially defined as a linear-quadratic mixture of the set of "original sources" $s_1(n), \ldots, s_M(n)$ contained in $s(n)$ may now be reinterpreted as a *linear* mixture of the set of "*extended* sources" contained in $\tilde{s}(n)$, with a corresponding "extended mixing matrix" $\tilde{A}$ instead of A. The reader may therefore wonder whether our investigation of linear-quadratic mixtures comes to an end thanks to this reinterpretation, because linear-quadratic mixtures are nothing but well-known linear mixtures, though of a modified set of source signals. The answer to that question is that linear-quadratic mixtures do require detailed additional investigations because, even if the nature of these mixtures may be connected to linear mixtures thanks to the above approach, other properties of the overall linear-quadratic configuration that we defined so far differ from the properties of the overall linear configurations that have been extensively studied in the literature. In particular:

- The components of the subvector $p(n)$ of $\tilde{s}(n)$ read as $s_j(n)s_k(n)$ and therefore have a functional dependence with respect to the components of its subvector $s(n)$, which read as $s_j(n)$. When considering stochastic original source signals $s_j(n)$, even if all these signals are mutually statistically independent, this independence property is thus not met by all components of the *extended* source vector $\tilde{s}(n)$ of the linearly-reformulated BSS problem corresponding to the mixing model (2.14). Therefore, one cannot straightforwardly apply to linear-quadratic mixtures the ICA methods, based on source mutual independence, which were historically the first class of BSS methods for linear mixtures and are still of major importance for such mixtures.
- Similarly, the concept of "determined mixtures" plays a major role for linear mixtures and yields additional constraints for linear-quadratic mixtures, as detailed below in Sect. 2.4. Before proceeding to that topic, we hereafter introduce a last expression for the linear-quadratic mixing model.

2.3 Overall Matrix Form of Mixing Model

All above mixing models were expressed for a single sample, with index n, of the source and observed vectors, $s(n)$ and $x(n)$. For linear mixtures, various BSS methods are expressed with respect to the complete available set of data, which corresponds to n ranging from 1 to N. To this end, one introduces the overall source and observation matrices

$$S = [s(1), \ldots, s(N)] \tag{2.15}$$

$$X = [x(1), \ldots, x(N)]. \tag{2.16}$$

Stacking column-wise the instances of (2.10) corresponding to all sample indices n then yields the overall matrix form of the linear mixing model, which reads

$$X = AS. \tag{2.17}$$

That approach then extends to the linear-quadratic single-sample model (2.14), including its bilinear version, and yields its overall matrix version

$$X = \tilde{A}\tilde{S}, \tag{2.18}$$

where

$$\tilde{S} = [\tilde{s}(1), \ldots, \tilde{s}(N)]. \tag{2.19}$$

2.4   Determined Mixtures of Original or Extended Sources

For linear mixtures, three subclasses of mixtures may be defined, depending on the numbers M and P of source and observed signals, respectively:

- The most standard subclass consists of determined mixtures, which correspond to $P = M$.
- Underdetermined mixtures correspond to $P < M$. Performing BSS with such mixtures is much harder than with determined ones, due to the reduced number of available observations, as explained in Sect. 3.1.
- In contrast, overdetermined mixtures correspond to $P > M$ and are much more easily handled than underdetermined ones. At least, one may just ignore $(P - M)$ observed signals and process the determined mixture of the remaining M observations.[5] Besides, for linear mixtures, one may make a better use of all available observations (especially if they contain additional noise) by using standard techniques, such as Principal Component Analysis [64], so as to derive M combinations of these observations.

Let us first extend the above discussion to the expressions (2.2) or (2.11) of linear-quadratic mixtures. In that approach, we will especially describe in this book how to handle the considered mixed signals $x_1(n), \ldots, x_P(n)$ in the case when they define a determined mixture of the original sources $s_1(n), \ldots, s_M(n)$, i.e., when $P = M$. Again, we will not consider the underdetermined version of that problem, i.e., when $P < M$, which is more difficult to handle. In contrast, overdetermined mixtures, i.e., when $P > M$, are not an issue because they may at least be transformed into determined ones by again ignoring $(P - M)$ observed signals. They may also be handled by methods based on Nonnegative Matrix Factorization, described further in this book.

Moreover, we explained above that a linear-quadratic mixture may also be reinterpreted as another, linear, mixture of an *extended* set of sources, as shown in (2.14). Based on the above-mentioned properties of linear mixtures with respect to the numbers of sources and mixtures, one may expect that this approach (2.14) to linear-quadratic mixtures opens a convenient alternative way to other types of BSS methods, provided the considered mixtures are determined (or overdetermined) with respect to the extended set of sources. This extended set of sources consists of the M original sources $s_1(n), \ldots, s_M(n)$ and, in the most general configuration, of the $\frac{M(M+1)}{2}$ source products $s_j(n)s_k(n)$ with $j = 1, \ldots, M$ and $k = j, \ldots, M$. Determined mixtures of extended sources then correspond to

$$P = \frac{M(M + 3)}{2}. \tag{2.20}$$

---

[5]As in the originally determined configuration, these observations are required to be linearly independent.

Similarly, for the subclass composed of bilinear mixtures, the determined configuration corresponds to

$$P = \frac{M(M+1)}{2}.$$

(2.21)

The second approach to linear-quadratic mixtures, based on the linear reformulation (2.14) is therefore much more demanding than the original one in terms of the number of observations P which is required to obtain determined mixtures (of the extended set of sources to be considered in that approach), for a given number M of sources. But it then allows one to derive simple separating system structures, as shown in the next chapter.

Chapter 3
Invertibility of Mixing Model, Separating Structures

In many BSS methods, once the considered class of mixing models has been selected, one defines the "structure" (as opposed to the associated adaptation procedure, considered later) of the separating system chosen for mainly estimating the source signals. The operation of this separating system is often defined by the equations that allow one to derive the output values of this system from its input values and from the considered values of its internal parameters, which are called the "separating parameters" hereafter. In Sects. 3.1–3.4, we define several such separating system structures for linear-quadratic mixtures. A different approach is then described in Sect. 3.5. All these structures open the way to the subsequent investigation of blind methods for adapting these separating systems, i.e., for selecting adequate *values* of their separating parameters.

3.1 A Linear Separating Structure for Determined Mixtures of Extended Sources

The simplest approach suggested by the discussion in Sect. 2.4 consists in reinterpreting a linear-quadratic mixture of the original sources as a linear mixture of the extended set of sources, defined by (2.14) or (2.18). In the case when this linear mixture is determined, as defined in (2.20), and when the square $P \times P$ matrix $\tilde{A}$ in this model is invertible, one may process the observed signals with the *linear* separating system structure, which has also been widely used in the literature for "originally linear mixtures," i.e., linear mixtures of original sources. In its simplest form, this structure consists of a square $P \times P$ separating matrix C, so that the output of the separating system for a given sample index n consists of the P-component vector

Y. Deville et al., *Nonlinear Blind Source Separation and Blind Mixture Identification*, SpringerBriefs in Electrical and Computer Engineering, https://doi.org/10.1007/978-3-030-64977-7_3

$$y(n) = Cx(n). \tag{3.1}$$

This separating system has a direct (or feedforward) structure, i.e., it directly derives its output vector $y(n)$ by combining (1) its input, which receives the observed vector $x(n)$, and (2) its internal separating matrix C. The adaptation of this separating system would then ideally consist in determining an estimate $\widehat{\tilde{A}}$ of $\tilde{A}$ (or of its inverse) and in setting

$$C = \left(\widehat{\tilde{A}}\right)^{-1}. \tag{3.2}$$

Equations (2.14) and (3.1)–(3.2) then yield

$$y(n) = \left(\widehat{\tilde{A}}\right)^{-1} \tilde{A}\tilde{s}(n), \tag{3.3}$$

i.e., $y(n) = \tilde{s}(n)$ up to estimation errors, where $\tilde{s}(n)$ is defined in (2.12). In particular, the M first components of $y(n)$ thus, respectively, provide estimates of $s_1(n), \ldots, s_M(n)$. The other outputs of that system then provide estimates of source products, which is of limited use and may even be an issue for some types of adaptation procedures discussed further in this book (see Sect. 4.1.1). A modified version of this structure, discussed in [42], therefore consists in only extracting the M source signals $s_1(n), \ldots, s_M(n)$ and thus consists of an $M \times P$ separating matrix C. Another version [48] splits the linear separating transform (3.1) into two successive transforms: the first one aims at removing the quadratic components from the observations, thus providing linear mixtures of the source signals, so that the second transform then performs conventional linear separation.

It should be noted that the mixing model (2.14) used in this approach, with a square $P \times P$ matrix $\tilde{A}$, is bijective, i.e., any observed vector $x(n)$ has exactly one inverse image with respect to this model, defined by the inverse of (2.14), which reads

$$\tilde{s}(n) = \tilde{A}^{-1}x(n). \tag{3.4}$$

This discussion also shows why this approach does not directly apply to *underdetermined* mixtures of the extended set of sources: for such mixtures, matrix $\tilde{A}$ is not invertible, so that the separating matrix cannot just be set to (3.2).

To summarize, this approach has the advantage of using a very simple separating system structure (namely, matrix C) but requires the number of observations P to be (much) larger than the number M of original sources:[1] its lower bound is defined by (2.20) for general linear-quadratic mixtures, which, e.g., yields $P = 5$

---

[1]Unless additional constraints on the source signals and/or mixing matrix are added, to make it possible to derive BSS methods suited to *underdetermined* mixtures, as was done in the literature for originally linear mixtures.

observations for $M = 2$ sources and $P = 65$ for $M = 10$. Even when restricting oneself to bilinear mixtures, the bound (2.21), e.g., yields $P = 3$ for $M = 2$ and $P = 55$ for $M = 10$.

In some applications, the above constraint is not an issue, because a very large number of mixtures of a limited number of sources may be obtained, i.e., the considered mixtures are largely overdetermined. This, e.g., concerns multispectral and hyperspectral images in remote sensing applications, especially when choosing to consider these data, so that each observation corresponds to a pixel (see the references provided in Chap. 1). In contrast, in various other applications such as the analysis of gas sensor array data [11], only a limited number of observations can easily be obtained. Beyond the above simple separating structure, other separating systems should therefore be developed in order to decrease the required number of observations. The target number is a priori $P = M$, since this situation can be handled for originally linear mixtures. For linear-quadratic mixtures too, this case $P = M$ (i.e., determined mixtures of original sources) can indeed be handled, by using the structures described hereafter. Of course, the latter structures also apply to the determined mixtures of extended sources (thus with $P > M$) considered above, at least by just ignoring $(P - M)$ observed signals. However, for such mixtures, the above linear separating structure is much more attractive than the nonlinear ones described below, due to their constraints that will appear in the following description.

3.2 Analytical Inversion of Determined Mixtures of Original Sources

Another attempt for building a separating system may be defined as follows. One first analytically solves the mixing equations (2.2) with respect to the sources $s_j(n)$ for the given observed values and mixing parameters, i.e., one derives the inverse of the direct model (2.2). In this analytical inverse model, one then replaces the mixing parameters a_{ij} and b_{ijk} respectively, by their estimates $\hat{a}_{ij}$ and $\hat{b}_{ijk}$, derived, e.g., by using one of the adaptation procedures described further in this book. The resulting equations provide estimates $\hat{s}_j(n)$ of the values $s_j(n)$ of the source signals, for each given vector $x(n)$ of observations and given values of $\hat{a}_{ij}$ and $\hat{b}_{ijk}$. These equations thus define a separating system.

This approach is applicable only if the analytical inverse of the considered mixing model can actually be derived. We hereafter first investigate this topic for one of the simplest versions of the linear-quadratic mixing model and then consider more general cases.

3.2.1 Bilinear Mixture of Two Original Sources

Let us first consider the mixing model defined by (2.4)–(2.5). These equations may be *analytically* solved, as explained in [57, 58]. For the given mixing parameter values a_{ij} and b_{ijk}, any observed vector $[x_1(n), x_2(n)]^T$ is thus shown to have two inverse images in the $[s_1(n), s_2(n)]^T$ plane. Note that the source-to-observation mapping considered here is thus non-bijective, unlike the mapping considered in Sect. 3.1. Each of these solutions $[s_1(n), s_2(n)]^T$ is defined by a closed-form nonlinear expression provided in [57, 58] and defines a specific separating system, which has a direct structure, i.e., which directly derives the source values (or their estimates $\hat{s}_j(n)$ in practice) from the observed vector and mixing parameters (or their estimates $\hat{a}_{ij}$ and $\hat{b}_{ijk}$ in practice).

These separating systems require some care however, because their indeterminacies depend on the sign of the Jacobian $J(n)$ of the normalized version[2] of the mixing equations for the considered observation sample $[x_1(n), x_2(n)]^T$ and mixing parameters [58]:

- If $J(n) < 0$, the separating system corresponding to the above first solution $[s_1(n), s_2(n)]^T$ restores the normalized sources without any indeterminacies, whereas the second system restores them up to permutation, scaling factors, and additive constants.
- In contrast, if $J(n) > 0$, the first separating system restores the normalized sources with the above-defined three indeterminacies, whereas the second system restores them without any indeterminacies.

Therefore, when using the same separating system, among the above two versions, to process a whole set of observed vectors, two situations may occur, depending on the considered data:

1. If $J(n)$ has the same sign for all considered data samples, the considered separating system restores the normalized sources with the same indeterminacies for all these samples. In particular, with respect to the permutation indeterminacy, this means that a given output of that system restores values corresponding to the *same* source for all observed vectors, as with conventional linear mixtures (and this restored source is here obtained possibly up to a scaling factor and an additive constant). This satisfactory situation is illustrated in Fig. 3.1.
2. In contrast, if the overall set of samples consists of a subset of samples such that $J(n) < 0$ and a subset such that $J(n) > 0$, the considered separating system restores the normalized sources without any indeterminacies for one of these subsets, and with the above indeterminacies for the other one. In particular, with respect to the permutation indeterminacy, this means that a given output of that system restores values corresponding to $s_1(n)$ for some observed vectors and to $s_2(n)$ for the others. This issue is illustrated in Fig. 3.2: the scatter plot of the

---

[2]This version uses the normalized sources defined in (3.7).

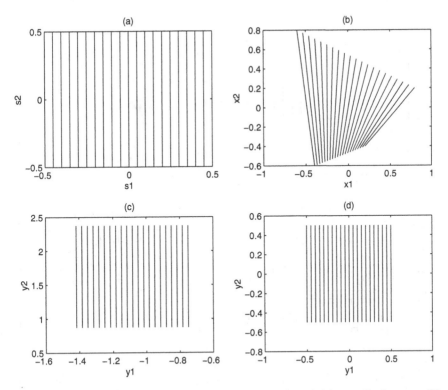

Fig. 3.1 Case when $J(n) > 0$ for all data samples. Scatter plots of: (**a**) normalized sources, (**b**) observed mixtures, (**c**) outputs of the first direct separating structure, and (**d**) outputs of the second direct separating structure

outputs of any given separating system (see Fig. 3.2c or d) does not look like the scatter plot of the normalized sources (see Fig. 3.2a), because that output scatter plot contains both types of points, i.e., with and without indeterminacies, especially with and without the permutation indeterminacy.

Using one of the above separating systems, a desirable situation is therefore when the source values are guaranteed to remain in bounded, suitable, intervals, so that $J(n)$ always has the same sign: each output of this separating system then only contains (estimated) samples of the same source.

3.2.2 More General Linear-Quadratic Mixtures

The target method in this Sect. 3.2, i.e., the analytical inversion of the mixing model, was shown above to be actually applicable to one of the simplest versions of the linear-quadratic model. Unfortunately, it cannot be developed for more complex

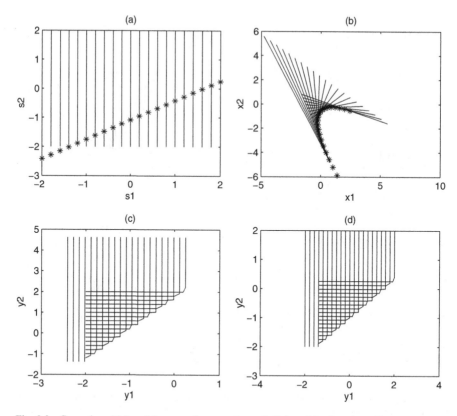

Fig. 3.2 Case when $J(n) > 0$ for some data samples and $J(n) < 0$ for the others. Scatter plots of: (**a**) normalized sources, (**b**) observed mixtures, (**c**) outputs of the first direct separating structure, and (**d**) outputs of the second direct separating structure. The stars in (**a**) and (**b**) correspond to $J(n) = 0$

versions of that model. For instance, even when restricting oneself to the three-source bilinear mixing model, one thus gets into the following dead end. In the resulting (normalized) three-source version of mixing model (2.2), with all $b_{ijk} = 0$ for $j = k$, one may remove $s_1(n)$ and $s_2(n)$ by combining the available three mixing equations. However, the resulting equation for $s_3(n)$ includes terms containing values of s_3^5, s_3^4, s_3^3, s_3^2, and s_3 and the square roots of polynomial functions of s_3 [59]. Such an equation cannot be analytically solved for $s_3(n)$. Therefore, to handle determined mixtures of original sources in more complex configurations than two-source bilinear mixtures, other separating system structures are required. Such structures are defined hereafter.

3.3 General-Purpose Numerical Inversion of Determined Mixtures of Original Sources

Let us again consider the mixing equations (2.2) with a_{ij}, b_{ijk} and $s_j(n)$ respectively, replaced by their estimates $\hat{a}_{ij}$, $\hat{b}_{ijk}$, and $\hat{s}_j(n)$. The method proposed at the beginning of Sect. 3.2 for deriving the estimated sources $\hat{s}_j(n)$ may be slightly changed, by *numerically* solving this modified version of Eqs. (2.2) with respect to the estimated sources $\hat{s}_j(n)$. In this framework, various general-purpose numerical methods from the literature may be used to solve these equations. For instance, the Newton–Raphson method is used in [8]. Although this may seem to provide a very simple and general approach for building separating systems for originally determined linear-quadratic mixing models, one should keep in mind that this approach has various potential issues:

- Even before focusing on a specific numerical method for solving nonlinear equations, one should take into account the intrinsic properties of these equations, i.e., of the considered mixing model. In the above discussion, we did not fully analyze the invertibility of the general linear-quadratic model (2.2), but we already mentioned the following results concerning the inverse images of a given observed vector: (1) for the basic, bilinear, two-source model, an observed vector has two inverse images, which do restore the sources, up to limited indeterminacies, and (2) even when only considering the slightly more complex bilinear, three-source model, the inverse images of an observed vector cannot be analytically derived. For the general linear-quadratic model, it therefore seems quite unlikely that one can analytically derive these inverse images. So, even if one succeeds in numerically deriving one of the solutions of the considered equations (up to numerical errors), e.g., by numerically minimizing an adequate cost function, it is not guaranteed that this solution provides the unmixed sources up to acceptable indeterminacies. Whereas this problem is due to the intrinsic nature of the mixing model, additional issues may then appear, due to the considered numerical method, as detailed hereafter.
- The source estimates $\hat{s}_j(n)$ may be defined as solutions of the above-mentioned equations. They therefore define one solution corresponding to the global minimum value, equal to zero, of specific cost functions that may be built for this type of equations. Numerical methods for minimizing such cost functions may reach this global minimum value, which is the case that we considered above. However, there is also a risk that they get stuck in a local minimum, where they do not provide unmixed sources. If using such a cost function, it is therefore highly desirable to analyze the positions and stability of all its minima.
- Beyond the stability of the equilibrium points of iterative optimization algorithms, another issue is the potentially chaotic behavior of these algorithms. This results from the fact that, unlike in conventional BSS, the mixing model and therefore the optimization algorithms considered here correspond to *nonlinear* mappings, which may entail chaotic behavior, as, e.g., detailed in [95].

Part of the above issues may be avoided with, or has been analyzed in more detail for, dedicated numerical methods, which are presented hereafter.

3.4 Dedicated Nonlinear Recurrent Neural Networks for Determined Mixtures of Original Sources

One of the earliest blind methods for separating source signals from their linear (memoryless) mixtures was proposed by Hérault and Jutten (see especially [56] and [68]). Its separating structure consists of a recurrent (or feedback) artificial neural network. Instead of that recurrent structure, most subsequently developed BSS methods for linear mixtures used the direct (or feedforward) structure defined by (3.1) that we considered in Sect. 3.1. This direct structure is preferred for linear mixtures, because it avoids the drawbacks of the recurrent one, namely an iterative computation of each output vector and potential instability issues. However, it was shown that recurrent structures may yield renewed interest when considering *nonlinear* mixtures (see especially [34]; see also previous related short communications, including [26, 57, 58]). This results from the fact that using recurrent separating system structures only requires one to know the analytical expression of the mixing model, whereas direct structures require one to know the analytical expression of the *inverse* of the mixing model. For linear mixtures, the expressions of both the mixing model and its inverse are known (both are linear). In contrast, when addressing nonlinear BSS for a given class of mixtures (with a determined mixture of the original sources), the analytical expression of the mixing model is known, but most often it cannot be analytically inverted so as to derive the analytical expression of the inverse of this mixing model (see, e.g., the above simple case dealing with the three-source bilinear model). Direct structures are then not applicable, whereas recurrent ones remain of interest, although they still have the limitations mentioned above for linear mixtures. We therefore describe these recurrent neural networks below, starting with a summary of their main principle of operation in the framework of linear mixtures and then showing how it extends to nonlinear ones.

3.4.1 Revisiting Linear Mixtures

For the sake of simplicity, we focus on the simplest version of determined linear (memoryless) mixtures, which concerns $P = 2$ mixtures of $M = 2$ sources. The linear mixing model (2.1) then explicitly reads

$$x_1(n) = a_{11}s_1(n) + a_{12}s_2(n) \tag{3.5}$$

$$x_2(n) = a_{21}s_1(n) + a_{22}s_2(n). \tag{3.6}$$

It may also be expressed with respect to the normalized source signals, which are defined as[3]

$$s_i'(n) = a_{ii} s_i(n) \qquad \forall i \in \{1, 2\} \tag{3.7}$$

and which are equal to the contribution of the source with index i in the observation with the same index i. Defining

$$L_{ij} = -\frac{a_{ij}}{a_{jj}} \qquad \forall i \neq j \in \{1, 2\} \tag{3.8}$$

(where we assume that all a_{jj} are non-zero), the above mixing equations (3.5)–(3.6) may be rewritten as

$$x_1(n) = s_1'(n) - L_{12} s_2'(n) \tag{3.9}$$

$$x_2(n) = -L_{21} s_1'(n) + s_2'(n). \tag{3.10}$$

As stated above, both source signals $s_i(n)$ may be extracted from these mixtures, up to the classical permutation and scale indeterminacies, by using the Hérault–Jutten recurrent structure. In [34], the operation of this structure was expressed in a way that had not previously been reported in the literature to our knowledge and that then made it possible to extend this structure to much more general mixtures, including the linear-quadratic and polynomial mixtures considered in this book. We hereafter summarize that analysis of the linear Hérault–Jutten neural network. For each sample index n, this structure receives the couple of observations $(x_1(n), x_2(n))$ and computes the values of its outputs y_i by performing a numerical recurrence. We denote as m the index associated with this recurrence and $y_i(m)$ the successive values of each output in this recurrence performed for sample n.[4] This recurrence reads

$$y_1(m + 1) = x_1(n) + l_{12} y_2(m) \tag{3.11}$$

$$y_2(m + 1) = x_2(n) + l_{21} y_1(m), \tag{3.12}$$

where l_{ij} are the weights of this neural network. The new values $y_i(m + 1)$ are then used as the input data of the next occurrence of the loop associated with this recurrence. The corresponding network may then also be considered as a looped structure (see Fig. 3.3), where the outputs $y_i(m)$ are fed back and combined with

---

[3] A dual normalization also exists: see [34].

[4] These successive output values therefore also depend on n. This index n is omitted in the notations $y_i(m)$, in order to improve readability and to focus on the recurrence on outputs for given input values $x_1(n)$ and $x_2(n)$.

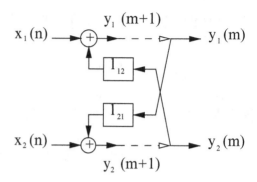

the inputs $x_i(n)$, to create new adder outputs $y_i(m+1)$, which then become the new outputs $y_i(m)$ for the next round through the loop.

Now consider the situation when the network weights are set to [34, 68]

$$l_{ij} = L_{ij} \qquad \forall\, i \neq j \in \{1, 2\}. \tag{3.13}$$

Let us assume that, at step m of the recurrence, we obtained

$$y_i(m) = s_i'(n) \qquad \forall\, i \in \{1, 2\}, \tag{3.14}$$

and let us consider the next occurrence of the recurrence loop. Figure 3.3 here becomes Fig. 3.4, and the network operates as follows. The outputs $y_i(m) = s_i'(n)$ are fed back, with the corresponding network weights $l_{ij} = L_{ij}$, and added to the observations. Each observation $x_i(n)$ defined by (3.9)–(3.10) consists of: (1) a "target term," which is the part of this observation that we would like to keep and which is equal to $s_i'(n)$ and (2) an "interfering term," which is the part of this observation that we would like to remove and which is equal to $-L_{ij}s_j'(n)$, with $j \neq i$. The major phenomenon is that, in the conditions considered here, each feedback term $l_{ij}y_j(m)$ becomes an exact "cancelling term," i.e., it exactly compensates for the above interfering term, so that the output of each adder becomes exactly equal to the target term $s_i'(n)$. We thus get

$$y_i(m + 1) = s_i'(n) \qquad \forall\, i \in \{1, 2\}, \tag{3.15}$$

i.e., condition (3.14) is still met after running one occurrence of the loop. It will therefore be met endlessly if the loop is executed again and again in these conditions. In more formal terms, this means that, in the conditions defined by (3.9)–(3.10) and (3.13), the point defined by (3.14) is a fixed point (i.e., an equilibrium point) of recurrence (3.11)–(3.12). This may be directly checked mathematically from all these equations, instead of considering Figs. 3.3 and 3.4.

We hereafter show that the above analysis extends to bilinear and more general nonlinear mixtures.

Fig. 3.4 Fixed point of loop of linear recurrent neural network

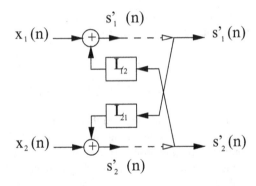

Fig. 3.5 Recurrent neural network for two-source bilinear mixing model: basic version, i.e., without self-feedback

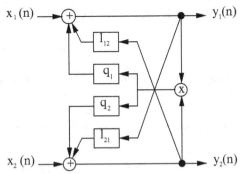

3.4.2 Bilinear Mixture of Two Original Sources

The two-source bilinear mixing model (2.4)–(2.5) may also be rewritten with respect to normalized sources, by still using (3.7)–(3.8) and now defining

$$Q_i = -\frac{b_i}{a_{11}a_{22}} \qquad \forall\, i \in \{1, 2\}, \tag{3.16}$$

where we still assume that all a_{jj} are non-zero. The above mixing equations (2.4)–(2.5) then read

$$x_1(n) = s_1'(n) - L_{12}s_2'(n) - Q_1 s_1'(n)s_2'(n) \tag{3.17}$$

$$x_2(n) = -L_{21}s_1'(n) + s_2'(n) - Q_2 s_1'(n)s_2'(n). \tag{3.18}$$

Each of these observations is therefore the sum of a target term equal to $s_i'(n)$ and of an interfering term that consists of all other components of $x_i(n)$ in (3.17)–(3.18).

The above discussion about linear mixtures therefore suggests that a first separating structure suited to the inversion of the bilinear mixing model considered here is the recurrent neural network shown in Fig. 3.5, whose operation for each observed vector $[x_1(n), x_2(n)]^T$ consists in computing its output vector $[y_1(n), y_2(n)]^T$ by means of the iterative algorithm[5]

$$y_1(m+1) = x_1(n) + l_{12}y_2(m) + q_1y_1(m)y_2(m) \qquad (3.19)$$

$$y_2(m+1) = x_2(n) + l_{21}y_1(m) + q_2y_1(m)y_2(m), \qquad (3.20)$$

where l_{ij} and q_i are the weights of this basic network. The feedback paths of this network aim at mimicking the structure of the opposite of the interfering terms contained in the observations, so as to cancel them in the network outputs for adequate values of the network parameters. More precisely, by using the same analysis procedure as in the above linear case, it may be easily shown that the network weights defined by (3.13) and

$$q_i = Q_i \qquad \forall i \in \{1, 2\} \qquad (3.21)$$

correspond to an equilibrium point of this network, which yields output signals again defined by (3.14). In other words, at this equilibrium point, the network succeeds in restoring the source signals (up to scale indeterminacies). This first analysis, however, leaves the following questions open:

1. Is this equilibrium point stable?
2. Does this network possess other equilibrium points, and if it does, do they restore the source signals and are they stable?

The stability of equilibrium points was studied in [34], thus showing that this basic network has a stable equilibrium point only for some values of the mixing coefficients and source signals. To solve this issue, an extended version of this network was proposed. Its structure is shown in Fig. 3.6, and its properties were studied in [34]. This network contains linear self-feedback loops (i.e., loops from each output to the input having the same index), with associated self-feedback weights l_{ii}. These parameters provide this network with additional flexibility, and the procedure proposed in [34] for automatically tuning them was shown to guarantee (local) convergence toward the above equilibrium point under mild conditions.

3.4.3 Other Linear-Quadratic Mixtures

The same type of approach as above may also be used to develop recurrent separating structures tailored to the two-source quadratic mixing model defined by (2.6)–(2.7). Various resulting recurrent neural networks were introduced in [34].

---

[5]Again, these successive output values therefore also depend on n, but this index n is omitted in the notations.

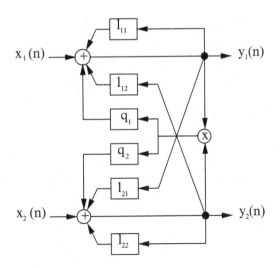

Fig. 3.6 Recurrent neural network for two-source bilinear mixing model: extended version, i.e., with self-feedback

The properties of this type of networks were then especially analyzed in [18, 19]. Beyond these specific mixing models, the above approach for building recurrent separating systems also applies to the most general form of the linear-quadratic mixing model (and to much wider classes of mixtures), as shown in [34]. That approach intrinsically guarantees that the case when the separating system outputs restore the source signals (up to limited indeterminacies) is an equilibrium point of these separating systems. However, one should then analyze these separating systems in more detail, to determine whether instability and chaotic behavior can be avoided at that desired equilibrium point, and whether other equilibrium points exist and how they behave, again with respect to instability and chaos. These properties may therefore limit the practical applicability of these neural networks for complicated mixing models.

As noted above, the equations used for computing the outputs of these recurrent networks are designed so as to correspond to the structure of the equations that define the mixing, i.e., *direct*, model: for example, compare (3.19)–(3.20) to (3.17)–(3.18). This yields the above-mentioned property: this approach does not require the analytical form of the *inverse* of the mixing model to be known, which is very attractive since the latter form is most often unknown for determined mixtures of original sources, as shown above. The inversion of the mixing model is here implicitly performed by the *recurrent* structure of the considered networks and the associated iterative algorithm.

3.5 Fitting the Direct Model

The "structure" considered here is again a "separating system" in the sense that it separates, i.e., restores, the source signals. However, it differs from the previous structures, because they essentially aim at separating the sources by transferring the observations through an estimate of the *inverse* of the mixing function, whereas the following structure aims at modeling that mixing, i.e., *direct*, function. This approach thus does not require one to know the analytical expression of the inverse model but only that of the direct model, which is a very attractive feature for nonlinear mixing models, as already mentioned above for some other approaches that also yield this feature.

More precisely, the approach considered here originates from methods mainly intended for (largely) overdetermined linear (memoryless) mixtures of nonnegative sources with nonnegative mixing coefficients, referred to as Nonnegative Matrix Factorization (NMF) methods. Various such linear methods were reported in the literature, especially based on the early works of Paatero et al. [88], who used the term "Positive Matrix Factorization," and those of Lee and Seung [72, 73], who introduced the term NMF. The latter methods start from the expression (2.17) of the actual data. They introduce adaptive matrices $\check{A}_\ell$ and $\check{S}_\ell$, which, respectively, aim at estimating A and S (up to permutation and scale indeterminacies). To this end, they adapt $\check{A}_\ell$ and $\check{S}_\ell$, essentially so that their product $\check{A}_\ell \check{S}_\ell$ fits the observed data matrix X, i.e., so that

$$X \approx \check{A}_\ell \check{S}_\ell. \tag{3.22}$$

The above approach extends to linear-quadratic mixtures, including bilinear ones (not necessarily with nonnegativity constraints for bilinear mixtures, as shown further in this book), by starting with their expression (2.18). In the resulting methods (see, e.g., [79, 80, 82] and subsequent works cited in Chap. 5), the separating system consists of two adaptive matrices $\check{A}$ and $\check{S}$ that, respectively, aim at estimating $\tilde{A}$ and $\tilde{S}$ (possibly up to some indeterminacies). To this end, they adapt $\check{A}$ and $\check{S}$, essentially so that their product $\check{A}\check{S}$ fits the observed data matrix X, i.e., so that

$$X \approx \check{A}\check{S}. \tag{3.23}$$

The adaptation procedures associated with this "separating system" and the separability properties of this type of approaches are detailed in Chap. 5.

Chapter 4
Independent Component Analysis and Bayesian Separation Methods

4.1 Methods for i.i.d. Sources

In the framework of statistical BSS methods, we first consider the approaches intended for the case when each source signal is stochastic and i.i.d, i.e., is a series of independent and identically distributed random variables. The BSS problem, initially expressed with respect to stochastic processes, may then be reformulated with respect to the corresponding random variables obtained when considering these processes at a single time. The BSS methods that have been proposed to handle this situation are presented hereafter.

4.1.1 Exploiting the Mutual Independence of the Outputs of a Separating System

The first historical class of BSS methods developed for determined linear memoryless mixtures consists of Independent Component Analysis (ICA) methods. In its strictest sense, ICA assumes the sources to be mutually statistically independent and it uses the independence of the outputs of a separating system,[1] thanks to the separation principle which consists in adapting the parameters of that system so that its outputs meet a criterion based on independence. More precisely:

---

[1] This system was originally [56, 68] the linear recurrent, or feedback, neural network described in Sect. 3.4.1. Most methods then used the system defined by (3.1), which may be considered as a direct, or feedforward, network.

© The Author(s), under exclusive license to Springer Nature Switzerland AG 2021
Y. Deville et al., *Nonlinear Blind Source Separation and Blind Mixture Identification*, SpringerBriefs in Electrical and Computer Engineering,
https://doi.org/10.1007/978-3-030-64977-7_4

- The early approach proposed by Hérault and Jutten (see especially [56] and [68]) uses an *approximation* of output independence. In its most general form, it consists of cancelling generalized cross-moments of the separating system outputs. In its simplest and most standard form, it cancels the (3,1) centered cross-moments of these outputs (it is well known that this version of the BSS problem cannot be solved by using only their second-order cross-moments).
- Subsequent works especially include the general method proposed by Comon in [23], which *completely* exploits output independence, by considering output mutual information.

Both above approaches were extended to linear-quadratic (memoryless) mixtures which are determined with respect to the original sources. These extensions were detailed for the two-source bilinear version of such mixtures, and both extensions use the associated basic recurrent network shown in Fig. 3.5. They are described hereafter.

4.1.1.1   A Moment-Based Method

The first proposed approach [57] may be seen as a bilinear extension of the Hérault-Jutten method, since it adapts the parameters of the above nonlinear recurrent neural network so as to achieve an approximation of output independence, more precisely so as to cancel the (3,1) or (2,1) partly centered output cross-moments:

- The separating network parameters l_{12} and l_{21}, corresponding to the linear terms in (3.19)–(3.20), are adapted so as to achieve

$$E\left\{y_1^3 \times [y_2 - E\{y_2\}]\right\} = 0 \tag{4.1}$$

$$E\left\{y_2^3 \times [y_1 - E\{y_1\}]\right\} = 0, \tag{4.2}$$

where $E\{.\}$ stands for expectation, whereas y_1 and y_2 are the random variables defined by the signals $y_1(n)$ and $y_2(n)$ at any given time n.
- The parameters q_1 and q_2, corresponding to the second-order terms in (3.19)–(3.20), are adapted so as to achieve

$$E\left\{y_1^2 \times [y_2 - E\{y_2\}]\right\} = 0 \tag{4.3}$$

$$E\left\{y_2^2 \times [y_1 - E\{y_1\}]\right\} = 0. \tag{4.4}$$

The separation conditions (4.1)–(4.2) are the same as those used in the original Hérault-Jutten network (except for signal centering), and conditions (4.3)–(4.4) are quite similar. The algorithm proposed in [57] for meeting the overall criterion (4.1)–(4.4) is therefore very similar to the algorithm used for the original Hérault-Jutten network.

4.1.1.2 An Approach Based on Mutual Information

For bilinear mixtures, the second approach based on output independence [86] is the bilinear extension of Comon's method in terms of separation criterion (but for a different separating system), since it adapts the above parameters so as to minimize, i.e., cancel, the mutual information of the network outputs, thus completely ensuring output independence. The selected network parameters are thus solutions of the separation criterion defined as

$$\left\{ \hat{l}_{12}, \hat{l}_{21}, \hat{q}_1, \hat{q}_2 \right\} = \underset{\left\{ l_{12}, l_{21}, q_1, q_2 \right\}}{\operatorname{argmin}} \; I(y_1, y_2), \qquad (4.5)$$

where $I(.)$ stands for mutual information. In [86], this mutual information is minimized by using a standard gradient descent algorithm. A similar approach, based on the separating structure of Sect. 3.3, was proposed in [8]. Besides, [47] proposed to replace the mutual information by the quadratic divergence between estimates of the joint probability density function (pdf) and the product of the marginal pdfs of the separated sources, as a cost function used for optimizing the weights of the above recurrent network. This optimization was performed by an evolutionary algorithm, namely opt-aiNet.

Still in the case of bilinear mixtures, let us now consider the situation when a higher number of observations are available, so that these mixtures are determined with respect to the *extended* sources. This opens the way to additional BSS methods, because the above recurrent structure and its issues may be avoided, especially by using the linear separating system described in Sect. 3.1, which is defined by a matrix C. More precisely, we here consider the modified version of this system restricted to M outputs. Each of these outputs aims at extracting one of the M original source signals $s_j(n)$, as opposed to the additional outputs that provide the source products $s_j(n)s_k(n)$ in the first version of this system described in Sect. 3.1. The system considered here is thus defined by a non-square $M \times P$ matrix C. This choice is motivated by the fact that we here focus on ICA-based BSS methods, i.e., methods which adapt the separating system parameters so as to enforce the mutual independence of all outputs of that system, and the source signals $s_j(n)$ to be thus extracted are indeed (assumed to be) mutually independent, whereas the source products $s_j(n)s_k(n)$ are not independent from them and from one another. Here again, output independence may be enforced by minimizing the mutual information of the separating system outputs. The selected separating matrix C is thus a solution of the separation criterion defined as

$$\hat{C} = \underset{C}{\operatorname{argmin}} \, I(y_1, \ldots, y_M). \qquad (4.6)$$

This approach was applied to $M = 2$ original sources in [42]. When using a standard gradient-based algorithm to optimize the above cost function, this algorithm got trapped in spurious minima. Therefore, the authors of [42] proposed a two-stage optimization procedure. In the first stage, they applied an evolutionary algorithm (opt-aiNet) to derive a solution which is coarse but close to the global minimum. They then refined that solution by performing a gradient-based descent.

4.1.2   Methods Focused on Estimating the Mixing Model

For determined linear memoryless mixtures, the term "Independent Component Analysis (ICA)" has also been used in a somewhat broader sense than above, by also including in the ICA field other methods which are not necessarily focused on the independence of the outputs of a separating system, but which again rely on the assumed mutual statistical independence of the source signals. Two such approaches and their extensions to linear-quadratic mixtures are considered hereafter.

4.1.2.1   The Maximum Likelihood Approach

The first considered approach is based on the maximum likelihood principle. For linear mixtures, it was especially introduced by Gaeta and Lacoume in [49], and by Pham and Garat in [89]. The maximum likelihood method may be defined as the estimation of the matrix A involved in (2.10), i.e., as the estimation of the parameters of the mixing (or direct) model, by selecting the parameter values which maximize the likelihood of the observed data. This presentation is, e.g., used in [15, 49, 89]. The maximum likelihood approach has also been defined as the estimation of the *inverse* A^{-1} of the above matrix A, i.e., as the estimation of the parameters of the matrix C of (3.1) to be used as the separating system, again by selecting the parameter values which maximize the likelihood of the observed data. This presentation is, e.g., used in [28, 64]. Both presentations may be equivalently used for linear mixtures, because the analytical expression of the inverse of the mixing model may be straightforwardly derived from the analytical expression of that mixing model, so that parametric models are available for both the direct and inverse transforms involved in this linear BSS problem.

In contrast, when extending the above maximum likelihood method to the general form of determined linear-quadratic mixtures of original sources, only the direct, i.e., mixing, model has a known analytical expression, as explained in Sect. 3.2. The maximum likelihood approach has therefore been applied to estimate the parameter values of that direct model, not to primarily identify the analytical expression and parameter values of the inverse model. That approach therefore first aims at achieving BMI. It was initially proposed for the two-original-source bilinear mixing model [58, 59],[2] then adapted to the two-original-source purely quadratic

---

[2]For this very specific mixing model, the analytical expression of the inverse of the mixing model is known, as explained in Sect. 3.2, and might therefore be used instead.

(i.e., without linear terms) mixing model [20] and subsequently extended to a configuration which also includes filtering effects [84].

More precisely, the investigations [58, 59] dealing with the two-original-source bilinear mixing model use a formulation of the mixing model to be estimated which is based on normalized source signals and that we here express as in (3.17)–(3.18), except that we replace the set $\{L_{12}, L_{21}, Q_1, Q_2\}$ of actual mixing parameters by the set $\{\check{L}_{12}, \check{L}_{21}, \check{Q}_1, \check{Q}_2\}$ of parameters to be estimated (see explanations in pp. 257–259 of [28] about the use of such different notations in general maximum likelihood approaches, to avoid ambiguity). The separation criterion used in the maximum likelihood approach then consists in selecting the values of these parameters which maximize the likelihood $\mathcal{L}$ of the observations, i.e.,

$$\left\{\hat{L}_{12}, \hat{L}_{21}, \hat{Q}_1, \hat{Q}_2\right\} = \underset{\left\{\check{L}_{12}, \check{L}_{21}, \check{Q}_1, \check{Q}_2\right\}}{\text{argmax}} \mathcal{L}. \tag{4.7}$$

The above likelihood function is defined as

$$\mathcal{L} = f_X(x_1(1), \ldots, x_P(1), \ldots, x_1(N), \ldots, x_P(N)), \tag{4.8}$$

where $f_X(.)$ is the joint pdf of all observations (here with $P = 2$), $x_j(n)$ is the value of the n-th sample of the j-th observation and N is the number of samples in the observations. The algorithm proposed in [58, 59] for maximizing the (logarithm of the) considered likelihood cost function is a gradient ascent algorithm. This approach requires one to compute the numerical values of the inverse images (scaled estimated sources) of the observations for the model defined by the version of (3.17)–(3.18) involving $\{\check{L}_{12}, \check{L}_{21}, \check{Q}_1, \check{Q}_2\}$. Therefore, although this maximum likelihood approach does not require the *analytical* expression of the inverse of the mixing model to be known, it requires the *numerical* version of that inverse model, as the methods for determined mixtures of the original sources based on the moments or mutual information of the output of the separating system described in Sect. 4.1.1. The maximum likelihood method considered here indeed has a close relationship with the above-defined approach based on output mutual information. That relationship was first discussed for linear memoryless mixtures and then extended in [37] to nonlinear mixtures, including linear-quadratic (memoryless) ones. In [58, 59], the numerical version of the inverse model is again obtained by using the recurrent networks defined in Sect. 3.4.

Once the mixing model has thus been identified (up to indeterminacies), this approach may be extended so as to achieve BSS, i.e., to restore the source signals in addition. To this end, one may use different separating structures defined in Chap. 3 for determined linear-quadratic mixtures of original sources. In particular, in the framework of the above-mentioned investigations [58, 59], it is then natural to use the recurrent networks defined in Sect. 3.4, as in the initial BMI phase.

Reference [48] also presents an approach related to likelihood maximization (and an associated semi-blind method), but dedicated to source signals which are Gaussian processes.

4.1.2.2   Methods Based on Cumulants and/or Moments

The second considered BMI approach consists in starting from the relationships which define the observed signals with respect to the source signals and mixing coefficients, and deriving resulting expressions of some cumulants or moments of the observed signals with respect to those of (power functions of) the source signals and to the mixing coefficients. Solving these equations for known (estimated) values of the observation cumulants or moments then especially yields the values of the mixing coefficients (up to some indeterminacies). This approach was first independently proposed in [25] and [78] for linear memoryless mixtures. It was then extended to two-original-source quadratic mixtures in [21]. That investigation uses a slightly different form of the considered mixing model, as compared with (2.6)–(2.7). This modified form amounts to requesting the coefficients $b_{111}, b_{122}, b_{211}$, and b_{222} of (2.6)–(2.7) to be nonnegative. Then denoting as $\beta_{111}, \beta_{122}, \beta_{211}$, and β_{222} the square roots of the coefficients $b_{111}, b_{122}, b_{211}$, and b_{222} of (2.6)–(2.7) and using $\beta_{112} = b_{112}, \beta_{212} = b_{212}$, the mixing model considered in [21] may be expressed as

$$x_1(n) = [\beta_{111}s_1(n)]^2 + \beta_{112}s_1(n)s_2(n) + [\beta_{122}s_2(n)]^2 \tag{4.9}$$

$$x_2(n) = [\beta_{211}s_1(n)]^2 + \beta_{212}s_1(n)s_2(n) + [\beta_{222}s_2(n)]^2. \tag{4.10}$$

Moreover, [21] considers a normalized version of this mixing model, derived similarly to (3.9)–(3.10) and (3.17)–(3.18). The normalized source signals are here defined as

$$s_i'(n) = \beta_{iii}s_i(n) \qquad \forall\, i \in \{1, 2\} \tag{4.11}$$

and the normalized quadratic mixing coefficients are

$$L_{ij} = -\frac{\beta_{ijj}^2}{\beta_{jjj}^2} \qquad \forall\, i \neq j \in \{1, 2\} \tag{4.12}$$

$$Q_i = -\frac{\beta_{i12}}{\beta_{111}\beta_{222}} \qquad \forall\, i \in \{1, 2\}, \tag{4.13}$$

where we again assume that the coefficients in the denominators of these expressions are non-zero. The mixing equations (4.9)–(4.10) may then be expressed as[3]

$$x_1(n) = \left[s_1'(n)\right]^2 - L_{12}\left[s_2'(n)\right]^2 - Q_1 s_1'(n)s_2'(n) \qquad (4.14)$$

$$x_2(n) = -L_{21}\left[s_1'(n)\right]^2 + \left[s_2'(n)\right]^2 - Q_2 s_1'(n)s_2'(n). \qquad (4.15)$$

In the following discussion, the prime signs used in the notations $s_i'(n)$ for *normalized* sources are omitted, for the sake of readability. Therefore, the notations $s_i(n)$ hereafter implicitly refer to these *normalized* sources.

The set of normalized mixing coefficients to be estimated is thus $\{L_{12}, L_{21}, Q_1, Q_2\}$. The method proposed in [21] to this end is based on cumulants and moments, which are statistical parameters defined e.g., in [28, 31, 64, 69, 83, 87]. We hereafter denote as $C[u_1, \ldots, u_K]$ the K-th order cross-cumulant of a set of K signals $\{u_1, \ldots, u_K\}$, where these signals are implicitly considered at a given time n, which has no influence on their cumulants since the considered signals are assumed to be (jointly) i.i.d. Moreover, when the same signal u_k is used several times, say r times, as an argument of a cumulant, shorter cumulant notations are used hereafter by replacing all these r arguments by the notation $(u_k)_r$. Thus, for example, $C[(s_1^2)_2]$ represents the second-order cumulant $C[s_1^2, s_1^2]$ of the squared source signal s_1^2. Similarly, $C[(x_1)_2, x_2]$ is the third-order cross-cumulant $C[x_1, x_1, x_2]$ which uses the observed signal x_1 twice as an argument and x_2 once. The method of [21] uses the first-order to fourth-order cumulants of the observations. It applies to the case when the source signals are mutually statistically independent and have symmetric pdfs (so that they are zero mean), which entails that some cumulants associated with these sources are zero. Using the mixing equations (4.14)–(4.15), the observation cumulants are shown in [21] to be expressed as detailed hereafter, with respect to (1) the (normalized) mixing coefficients and to (2) cumulants of power functions of the (normalized) sources or eventually moments of these sources.

The first-order cumulants of the observations read

$$\begin{cases} C[x_1] = E\{s_1^2\} - L_{12}E\{s_2^2\} \\ C[x_2] = -L_{21}E\{s_1^2\} + E\{s_2^2\} \end{cases} . \qquad (4.16)$$

The second-order cumulants of the observations are first derived with respect to cumulants of power functions of the source signals. The above-defined assumptions concerning the sources yield

---

[3]The mixing equations (4.14)–(4.15) obtained here are similar to the Eqs. (3.17)–(3.18) derived above for the bilinear mixing model, thanks to the use of the notations (4.12)–(4.13). However, it should be clear that this is here obtained by using the notations L_{ij} for coefficients associated with *quadratic* (auto-)terms, whereas in the bilinear model (3.17)–(3.18) they were used for coefficients associated with *linear* terms, which was more coherent with these notations L_{ij}.

$$\begin{cases} C[(x_1)_2] = C\big[(s_1^2)_2\big] + L_{12}^2 C\big[(s_2^2)_2\big] + Q_1^2 C[(s_1 s_2)_2] \\ C[(x_2)_2] = L_{21}^2 C\big[(s_1^2)_2\big] + C\big[(s_2^2)_2\big] + Q_2^2 C[(s_1 s_2)_2] \\ C[x_1, x_2] = -L_{21} C\big[(s_1^2)_2\big] - L_{12} C\big[(s_2^2)_2\big] + Q_1 Q_2 C[(s_1 s_2)_2] \end{cases} \tag{4.17}$$

and the above cumulants related to the sources may then be expressed as follows with respect to source moments:

$$\begin{cases} C\big[(s_1^2)_2\big] = E\{s_1^4\} - E\{s_1^2\}^2 \\ C\big[(s_2^2)_2\big] = E\{s_2^4\} - E\{s_2^2\}^2 \\ C[(s_1 s_2)_2] = E\{s_1^2\} E\{s_2^2\}. \end{cases} \tag{4.18}$$

Similarly, the third-order observation cumulants read

$$\begin{cases} C[(x_1)_3] = C\big[(s_1^2)_3\big] - L_{12}^3 C\big[(s_2^2)_3\big] \\ \qquad\qquad + 3Q_1^2 C\big[s_1^2, (s_1 s_2)_2\big] - 3L_{12} Q_1^2 C\big[s_2^2, (s_1 s_2)_2\big] \\ C[(x_2)_3] = C\big[(s_2^2)_3\big] - L_{21}^3 C\big[(s_1^2)_3\big] \\ \qquad\qquad - 3L_{21} Q_2^2 C\big[s_1^2, (s_1 s_2)_2\big] + 3Q_2^2 C\big[s_2^2, (s_1 s_2)_2\big] \\ C[(x_1)_2, x_2] = -L_{21} C\big[(s_1^2)_3\big] + L_{12}^2 C\big[(s_2^2)_3\big] \\ \qquad\qquad + (2Q_1 Q_2 - L_{21} Q_1^2) C\big[s_1^2, (s_1 s_2)_2\big] \\ \qquad\qquad + (Q_1^2 - 2L_{12} Q_1 Q_2) C\big[s_2^2, (s_1 s_2)_2\big] \\ C[x_1, (x_2)_2] = L_{21}^2 C\big[(s_1^2)_3\big] - L_{12} C\big[(s_2^2)_3\big] \\ \qquad\qquad + (Q_2^2 - 2L_{21} Q_1 Q_2) C\big[s_1^2, (s_1 s_2)_2\big] \\ \qquad\qquad + (2Q_1 Q_2 - L_{12} Q_2^2) C\big[s_2^2, (s_1 s_2)_2\big] \end{cases} \tag{4.19}$$

and the cumulants related to the sources which appear in the above expressions may then be expressed as

$$\begin{cases} C\big[(s_1^2)_3\big] = E\{s_1^6\} - 3E\{s_1^4\} E\{s_1^2\} + 2E\{s_1^2\}^3 \\ C\big[(s_2^2)_3\big] = E\{s_2^6\} - 3E\{s_2^4\} E\{s_2^2\} + 2E\{s_2^2\}^3 \\ C\big[s_1^2, (s_1 s_2)_2\big] = E\{s_2^2\} E\{s_1^4\} - E\{s_2^2\} E\{s_1^2\}^2 \\ C\big[s_2^2, (s_1 s_2)_2\big] = E\{s_1^2\} E\{s_2^4\} - E\{s_1^2\} E\{s_2^2\}^2. \end{cases} \tag{4.20}$$

Finally, the fourth-order observation cumulants are

$$
\begin{cases}
C[(x_1)_4] = C\Big[\big(s_1^2\big)_4\Big] + L_{12}^4 C\Big[\big(s_2^2\big)_4\Big] + Q_1^4 C\Big[\big(s_1 s_2\big)_4\Big] \\
\qquad + 6Q_1^2 C\Big[\big(s_1^2\big)_2, (s_1 s_2)_2\Big] + 6L_{12}^2 Q_1^2 C\Big[\big(s_2^2\big)_2, (s_1 s_2)_2\Big] \\
\qquad - 12 L_{12} Q_1^2 C\Big[s_1^2, s_2^2, (s_1 s_2)_2\Big] \\[4pt]
C[(x_2)_4] = L_{21}^4 C\Big[\big(s_1^2\big)_4\Big] + C\Big[\big(s_2^2\big)_4\Big] + Q_2^4 C\Big[\big(s_1 s_2\big)_4\Big] \\
\qquad + 6L_{21}^2 Q_2^2 C\Big[\big(s_1^2\big)_2, \big(s_1 s_2\big)_2\Big] + 6Q_2^2 C\Big[\big(s_2^2\big)_2, \big(s_1 s_2\big)_2\Big] \\
\qquad - 12 L_{21} Q_2^2 C\Big[s_1^2, s_2^2, (s_1 s_2)_2\Big] \\[4pt]
C[(x_1), (x_2)_3] = -L_{21}^3 C\Big[\big(s_1^2\big)_4\Big] - L_{12} C\Big[\big(s_2^2\big)_4\Big] + Q_1 Q_2^3 C[(s_1 s_2)_4] \\
\qquad + \Big(-3Q_2^2 L_{21} + 3Q_1 Q_2 L_{21}^2\Big) C\Big[\big(s_1^2\big)_2, (s_1 s_2)_2\Big] \\
\qquad + \Big(-3L_{12} Q_2^2 + 3Q_1 Q_2\Big) C\Big[\big(s_2^2\big)_2, (s_1 s_2)_2\Big] \\
\qquad + \Big(3Q_2^2 - 6L_{21} Q_2 Q_1 + 3L_{12} L_{21} Q_2^2\Big) C\Big[s_1^2, s_2^2, (s_1 s_2)_2\Big] \\[4pt]
C[(x_2), (x_1)_3] = -L_{21} C\Big[\big(s_1^2\big)_4\Big] - L_{12}^3 C\Big[\big(s_2^2\big)_4\Big] + Q_2 Q_1^3 C[(s_1 s_2)_4] \\
\qquad + \Big(-3L_{21} Q_1^2 + 3Q_2 Q_1\Big) C\Big[\big(s_1^2\big)_2, (s_1 s_2)_2\Big] \\
\qquad + \Big(-3L_{12} Q_1^2 + 3Q_2 Q_1 L_{12}^2\Big) C\Big[\big(s_2^2\big)_2, (s_1 s_2)_2\Big] \\
\qquad + \Big(3Q_1^2 - 6L_{12} Q_2 Q_1 + 3L_{12} L_{21} Q_1^2\Big) C\Big[s_1^2, s_2^2, (s_1 s_2)_2\Big] \\[4pt]
C[(x_1)_2, (x_2)_2] = L_{21}^2 C\Big[\big(s_1^2\big)_4\Big] + L_{12}^2 C\Big[\big(s_2^2\big)_4\Big] + Q_2^2 Q_1^2 C[(s_1 s_2)_4] \\
\qquad + \Big(L_{21}^2 Q_1^2 + Q_2^2 - 4Q_1 Q_2 L_{21}\Big) C\Big[\big(s_1^2\big)_2, (s_1 s_2)_2\Big] \\
\qquad + \Big(L_{12}^2 Q_2^2 + Q_1^2 - 4L_{12} Q_1 Q_2\Big) C\Big[\big(s_2^2\big)_2, (s_1 s_2)_2\Big] \\
\qquad + \Big(4Q_1 Q_2 - 2L_{12} Q_2^2 - 2L_{21} Q_1^2 \\
\qquad\qquad + 4Q_1 Q_2 L_{12} L_{21}\Big) C\Big[s_1^2, s_2^2, (s_1 s_2)_2\Big]
\end{cases}
$$

$$(4.21)$$

and the corresponding cumulants related to the sources read

$$\begin{cases} C\left[\left(s_1^2\right)_4\right] = E\left\{s_1^8\right\} - 4E\left\{s_1^6\right\}E\left\{s_1^2\right\} - 6E\left\{s_1^2\right\}^4 \\ \qquad\qquad -3E\left\{s_1^4\right\}^2 + 12E\left\{s_1^4\right\}E\left\{s_1^2\right\}^2 \\ C\left[\left(s_2^2\right)_4\right] = E\left\{s_2^8\right\} - 4E\left\{s_2^6\right\}E\left\{s_2^2\right\} - 6E\left\{s_2^2\right\}^4 \\ \qquad\qquad -3E\left\{s_2^4\right\}^2 + 12E\left\{s_2^4\right\}E\left\{s_2^2\right\}^2 \\ C[(s_1 s_2)_4] = E\left\{s_1^4\right\}E\left\{s_2^4\right\} - 3E\left\{s_1^2\right\}^2 E\left\{s_2^2\right\}^2 \\ C\left[\left(s_1^2\right)_2, (s_1 s_2)_2\right] = E\left\{s_1^6\right\}E\left\{s_2^2\right\} - E\left\{s_1^2\right\}E\left\{s_2^2\right\}\left(3E\left\{s_1^4\right\} - 2E\left\{s_1^2\right\}^2\right) \\ C\left[\left(s_2^2\right)_2, (s_1 s_2)_2\right] = E\left\{s_2^6\right\}E\left\{s_1^2\right\} - E\left\{s_2^2\right\}E\left\{s_1^2\right\}\left(3E\left\{s_2^4\right\} - 2E\left\{s_2^2\right\}^2\right) \\ C\left[s_1^2, s_2^2, (s_1 s_2)_2\right] = E\left\{s_1^4\right\}E\left\{s_2^4\right\} - E\left\{s_1^2\right\}^2 E\left\{s_2^4\right\} \\ \qquad\qquad -E\left\{s_2^2\right\}^2 E\left\{s_1^4\right\} + E\left\{s_1^2\right\}^2 E\left\{s_2^2\right\}^2. \end{cases}$$

$$(4.22)$$

Equation (4.16), (4.17), (4.19), and (4.21) thus yield a set of 14 equations, where the left-hand terms are known, i.e., estimated in practice from the observed values. Combining these equations with (4.18), (4.20), and (4.22) yields a set of 14 equations which contain 12 unknowns, namely L_{12}, L_{21}, Q_1, Q_2, $E\{s_1^2\}$, $E\{s_1^4\}$, $E\{s_1^6\}$, $E\{s_1^8\}$, $E\{s_2^2\}$, $E\{s_2^4\}$, $E\{s_2^6\}$, and $E\{s_2^8\}$. The criterion used in this BMI method therefore consists of solving the above equations, especially in order to derive their subset of unknowns $\{L_{12}, L_{21}, Q_1, Q_2\}$. It should be noted that these equations are nonlinear with respect to $\{L_{12}, L_{21}, Q_1, Q_2\}$ and much more complex than in the original version of that method intended for linear memoryless mixtures. Given the above separation criterion, various associated algorithms may be considered. In [21], the above equations are solved numerically, by using the Levenberg-Marquardt or Gauss-Newton algorithms, as implemented in the fsolve function of the Matlab software.

Here again, once the (normalized) mixing model has thus been identified, this approach may be extended so as to restore the source signals in addition. To this end, one may again use different separating structures defined in Chap. 3, by downloading the estimated mixing parameters $\{\hat{L}_{12}, \hat{L}_{21}, \hat{Q}_1, \hat{Q}_2\}$ into such a separating structure (provided it applies to the considered source properties).

As stated above, in this book we mainly consider real-valued signals. In addition, we here briefly mention that a quite different BMI method was proposed in [70, 71] for complex-valued circular sources, also using cumulants.

4.1.3 Jointly Estimating the Sources and Mixing Model

The above-defined methods initially put more emphasis on *one* of the two types of unknowns of the BSS/BMI problem, namely on the source values or on the mixing model parameters. In contrast, other methods for linear-quadratic mixtures consider *both* types of unknowns in the same framework and aim at jointly estimating them. This joint approach especially includes Bayesian methods[4] which were, e.g., proposed for handling the blind[5] separation of i.i.d. mutually statistically independent sources mixed according to the bilinear model. This approach was introduced in [40, 41] and is presented in more detail in [43].

Unlike approaches described in the previous sections, Bayesian methods do not explicitly use a separating system as defined in Chap. 3, and thus avoid the associated potential issues, mainly for determined mixtures of original sources. Instead, Bayesian methods may be considered as a tool for modelling the considered data, namely the source values and mixing parameters (see [39, p. 109–110]). These two types of unknowns are gathered in a vector θ. Moreover, the Bayesian approach makes it possible to take into account prior knowledge about the sources and mixing model. This knowledge is expressed in terms of the prior pdf $f(\theta)$ of the above unknowns. More precisely, this pdf typically belongs to a given class of functions with some unknown parameters. These additional unknown parameters are also included in the vector θ of parameters to be estimated. In the approach proposed in [40, 41, 43], the sources have truncated Gaussian distributions, defined as

$$f\left(s_j(n)|\mu_j, p_j, s_j^{min}, s_j^{max}\right) = \frac{\sqrt{\frac{p_j}{2\pi}} \exp\left(-\frac{p_j}{2}\left(s_j(n)-\mu_j\right)^2\right) \mathbb{1}_{\left[s_j^{min},s_j^{max}\right]}(s_j(n))}{\Phi\left(\sqrt{p_j}\left(s_j^{max}-\mu_j\right)\right) - \Phi\left(\sqrt{p_j}\left(s_j^{min}-\mu_j\right)\right)},$$

$$(4.23)$$

where μ_j and p_j are unknown parameters to be estimated and are therefore inserted in θ, whereas the minimum and maximum source values s_j^{min} and s_j^{max} are fixed and set by the user. $\Phi(.)$ is the cumulative distribution function of the standard Gaussian distribution. The other unknown parameters included in the vector θ of parameters to be estimated are the variances of the noise components in each observation.

Once the above data model has been selected, the Bayesian approach contains two additional stages. The first one consists in combining the above prior information and the available observations, so as to derive the posterior pdf $f(\theta|X)$ of θ. Bayes' theorem then yields

$$f(\theta|X) = \frac{f(X|\theta)f(\theta)}{f(X)} \qquad (4.24)$$

$$\propto f(X|\theta)f(\theta), \qquad (4.25)$$

where $f(X|\theta)$ is the likelihood and $\propto$ stands for "proportional to."

---

[4]In the general framework of BSS/BMI, Bayesian methods may also be used to estimate only one of the above types of unknowns: see, e.g., p. 471 of [85].

[5]For linear-quadratic mixing models, and especially bilinear ones, various other investigations were also reported concerning the use of Bayesian methods, but for the non-blind (or supervised) configuration, i.e., when the mixing model parameters are known and only the sources are to be estimated (see, e.g., [54]), which is out of the scope of this book. Other types of non-blind methods for such mixtures are reported, e.g., in [76].

The above posterior pdf is then used to make inferences about θ, so as to eventually derive an estimate of θ. This may be achieved in different ways in the Bayesian framework (see, e.g., pp. 493–494 of [85]). The approach proposed in [40, 41, 43] uses the Bayesian minimum mean square error (MMSE) estimator (also known as the conditional mean estimator), which reads

$$\theta_{MMSE} = \int \theta f(\theta|X) d\theta. \tag{4.26}$$

The above integral cannot be easily calculated in the considered configuration. Therefore, in practice, it is replaced by the approximation

$$\hat{\theta} = \frac{1}{K} \sum_{k=1}^{K} \theta(k), \tag{4.27}$$

where $\{\theta(1), \ldots, \theta(K)\}$ is a set of samples of θ drawn from $f(\theta|X)$. In [40, 41, 43], these samples are obtained by using the Gibbs sampler. A similar approach was reported in [77] for a linear-quadratic mixing model.

Besides, a method based on the Maximum A Posteriori (MAP) was proposed in [66] for a remote sensing application. The priors were chosen accordingly, as follows. The sources are reflectance spectra of physical pure materials, with no prior information (except that they range from 0 to 1), so that a non-informative uniform prior was used for their distribution. For each image pixel considered in this application, the linear mixing coefficients represent the fractions of surface associated with each pure material in that pixel, so that the sum of these mixing coefficients is equal to one. A Dirichlet distribution was therefore used for them. Moreover, a half-normal distribution was used for the quadratic mixing coefficients, based on a numerical analysis of the properties of the data faced in the target application.

It should be noted that Bayesian BSS/BMI methods differ from those described in the previous sections of this book in the sense that only those previous methods *require* the source signals to be mutually statistically independent, at least in their reported versions. Bayesian methods can benefit from this assumption because the data model thus becomes simpler, but these methods are also applicable to situations when the sources have known dependence (by incorporating this information in prior distributions). However, inference is more complicated in this case and there is no guarantee that Bayesian methods then yield a unique data representation (see [39, pp. 96 and 108–109]).

Besides, it should be kept in mind that, in the framework of linear BSS/BMI, the development of the use of Bayesian methods has been limited by the complexity of their implementation and their high computational cost, as compared with the most popular ICA-related methods. These features of Bayesian methods might also limit their expansion for linear-quadratic mixtures.

4.2   Methods for Non-i.i.d. Sources

Beyond the above configurations, other linear-quadratic BSS methods may be developed by considering non-i.i.d. random source signals and exploiting their autocorrelation (when each source is not independently distributed for different samples n) and/or their non-stationarity (when each source is not identically distributed for different samples). These two approaches were highlighted in [16] for linear mixtures, in addition to the case of i.i.d. sources. They were used as follows for linear-quadratic mixtures.

4.2.1   Methods Focused on Estimating the Sources or Mixing Model

4.2.1.1   The Maximum Likelihood Approach

The extension of the above maximum likelihood approach to linear-quadratic mixtures of non-i.i.d. sources was reported in [60], where different cases are considered, depending whether the sources are autocorrelated (they are q-th order Markov processes) and/or non-stationary. This method may also be seen as the linear-quadratic extension of several investigations which were previously reported for *linear* mixtures of non-i.i.d. sources, namely [61] for stationary autocorrelated (Markovian) one-dimensional signals, [52] for possibly non-stationary autocorrelated (Markovian) one-dimensional signals, and [53] for stationary or non-stationary autocorrelated (Markovian) two-dimensional signals.

4.2.1.2   Moment-Based Methods

A BMI method for linear-quadratic mixtures of complex-valued, circular, stationary, autocorrelated, and mutually independent source signals was also proposed in [1], using a joint diagonalization of a set of observation correlation matrices. Using similar tools, a method for extracting source products was also presented in [50] for uncorrelated sources with distinct autocorrelation functions, mixed according to a model which only contains second-order auto-terms and cross-terms. Besides, a method for bilinear mixtures of autocorrelated and mutually independent sources was proposed in [67]. This method uses second-order statistics to extract the (centered version of the) source signals with an approach which may be seen as an extension of the well-know SOBI method [12] intended for linear instantaneous mixtures. A subsequent step of the method of [67] then makes it possible to estimate the mixing parameters.

4.2.2 Bayesian Method

In [40, 41, 43], the Bayesian method for i.i.d. sources described in Sect. 4.1 was also extended to non-i.i.d. sources. This is achieved by replacing the parameter μ_j in Eq. (4.23), which defines the distribution of sample n of source s_j, by the previous value of that source, that is $s_j(n-1)$. The sources are then first-order Markovian processes. The resulting Bayesian method is then derived almost in the same way as in the i.i.d. case, but there is no need to estimate μ_j in the Markovian case.

Chapter 5
Matrix Factorization Methods

5.1 General Features and Separation Principle

The class of BSS methods described in this chapter has the following features with respect to the procedure defined in Chap. 1. The considered mixing model is again the linear-quadratic model or the somewhat more specific bilinear model, which were both defined in Chap. 2 and which may be summarized by the mixing equation

$$X = \tilde{A}\tilde{S}. \tag{5.1}$$

These mixing models lead to the separating systems which were reviewed in Chap. 3. Among these systems, the BSS methods considered here only use the model defined in Sect. 3.5 and associated variants. As explained in that Sect. 3.5, the plain version of that separating system consists of adaptive matrices $\check{A}$ and $\check{S}$ which, respectively, aim at estimating $\tilde{A}$ and $\tilde{S}$. The associated variants are derived from this system by introducing some of the following types of constraints:

- Structural constraints, i.e., constraints on the structure of the separating system, in terms of which of its adaptive parameters (i.e., which of the elements of $\check{A}$ and $\check{S}$) are the so-called slave variables [30], i.e., variables whose values are derived from other, free, variables, which are called "master variables."
- Constraints on the domains of values in which the above variables are forced to stay. The most standard constraint, e.g., motivated by the remote sensing applications mentioned in Chap. 1, consists in requesting the values of all adaptive parameters associated with sources and mixing coefficients to be nonnegative. More specific constraints, e.g., appear in urban remote sensing applications [81, 82], where quadratic mixing coefficients were shown to typically belong to the interval $[0, 0.5]$, and the associated adaptive parameters may therefore be constrained to stay in this interval.

© The Author(s), under exclusive license to Springer Nature Switzerland AG 2021
Y. Deville et al., *Nonlinear Blind Source Separation and Blind Mixture Identification*, SpringerBriefs in Electrical and Computer Engineering, https://doi.org/10.1007/978-3-030-64977-7_5

- The so-called sum-to-one constraint, which is detailed further in this section (see (5.4)) and is also, e.g., considered in remote sensing applications.

This chapter provides a survey of the methods (in terms of separation principle, criterion, and algorithm) that have been proposed for adapting the tunable parameters of these separating systems, depending on the considered constraints. Each of its sections corresponds to one case in terms of structural constraints. As outlined in Sect. 3.5, the separation principle used in all these methods is based on adapting $\check{A}$ and $\check{S}$ so that their product $\check{A}\check{S}$ fits the observed data matrix X, in order to ideally achieve $\check{A}\check{S} = X$. These methods may be considered as data representation tools, which model the considered data, namely the source values and mixing parameters. They share this feature with the Bayesian methods described in Chap. 4. However, whereas the latter methods are based on a probabilistic framework, the approaches presented below use deterministic representations and therefore other separation criteria than the Bayesian methods.

5.2   Methods Without Structural Constraints

A first approach consists in considering the mixing model (5.1) without taking into account the internal structure of $\tilde{S}$. The observed signals contained in X are then seen as linear mixtures of the overall set of signals (extended sources) contained in $\tilde{S}$, as already noted in Sect. 2.2. They may therefore be estimated by using the matrix factorization methods which were reported for *linear* mixtures (this also yields the extended mixing matrix $\tilde{A}$, thus also achieving BMI). These methods were outlined in Sect. 3.5 of this book and presented in more detail e.g., in [22]. The only difference here is that, for a given number M of original sources, these methods are typically applied by setting the number of columns of $\check{A}$ and the number of rows of $\check{S}$ to the (known or estimated) number of *extended* sources, that is $\frac{M(M+3)}{2}$ for linear-quadratic mixtures and $\frac{M(M+1)}{2}$ for bilinear ones, as explained in Sect. 2.4.

Moreover, it should be remembered that, to apply the above linear matrix factorization methods to arbitrary[1] sources which are actually mixed according to the *linear* mixing model, additional constraints must be set on the source values and mixing coefficients. This may be explained as follows, considering the matrix form (2.17) of the linear mixing model and associated tunable matrices $\check{A}_\ell$ and $\check{S}_\ell$, respectively, used to estimate A and S. When each of the rows of $\check{S}_\ell$ is an arbitrary linear combination of the actual source vectors which form the rows of S, the rows of $\check{A}_\ell \check{S}_\ell$ are linear combinations of the actual source vectors, as the rows of X. If the rows of $\check{S}_\ell$ span (at least) the same subspace as the rows of X, the coefficients in $\check{A}_\ell$ can then always be selected so that $\check{A}_\ell \check{S}_\ell = X$ exactly.

---

[1]We mean sources which are mutually unconstrained, unlike the extended set of sources involved in the linear-quadratic mixtures addressed in this book.

This linear version of the (unconstrained) Matrix Factorization separation principle therefore does not allow one to retrieve the sources: $\check{S}_\ell$ may thus yield any linear *mixture* of these sources. If trying to use the (linear) Matrix Factorization principle for the linear mixing model, the data must therefore be further constrained. As stated above, the considered constraint most often consists in requiring the (actual and estimated) source and mixing coefficients to be nonnegative, thus leading to (linear) Nonnegative Matrix Factorization (NMF) methods [22], although this nonnegativity constraint has been shown not to be sufficient for ensuring the uniqueness of the linear NMF decomposition.

Now, when applying the above linear NMF approach to the extended set of source values and associated coefficients contained in $\tilde{A}$ and $\tilde{S}$, various methods may be defined for measuring how well the associated adaptive matrices $\check{A}$ and $\check{S}$ yield a product $\check{A}\check{S}$ which fits the observed data X. We hereafter consider the most standard method, where the "distance" between X and $\check{A}\check{S}$ is defined as the Frobenius norm of the difference between X and $\check{A}\check{S}$, denoted as $||X - \check{A}\check{S}||_F$. The separation criterion of this Linear Extended Nonnegative Matrix Factorization (LE-NMF) method consists in selecting values of $\check{A}$ and $\check{S}$ which minimize the above norm or its square,[2] subject to the above-defined nonnegativity constraints, that is

$$\left\{\widehat{\tilde{A}}, \widehat{\tilde{S}}\right\}_{LE-NMF} = \underset{\{\check{A},\check{S}\}}{\operatorname{argmin}} \frac{1}{2}||X - \check{A}\check{S}||_F^2 \qquad \text{subject to } \check{A} \geq 0, \check{S} \geq 0. \quad (5.2)$$

Associated algorithms for optimizing $\check{A}$ and $\check{S}$ are then obtained by just applying algorithms for linear mixtures (with the above-defined dimensions of $\check{A}$ and $\check{S}$ corresponding to the number of *extended* sources). We stress that these algorithms are not modified as compared with the standard linear case only because the internal structure of $\tilde{S}$ (defined by (2.12) and (2.19)) is not taken into account, i.e., all elements of $\check{S}$ are here considered as independent[3] variables, although they aim at estimating elements of $\tilde{S}$ which do possess functional dependencies. Disregarding these dependencies leads to simpler algorithms but is likely to yield lower performance, because it does not exploit some available knowledge about the considered data structure. This sub-optimality was experimentally confirmed, e.g., in [82]. This LE-NMF method was also considered in [79].

---

[2]The square of this norm is more easily handled when then considering algorithms which use the gradient of this cost function. A factor of 1/2 is then inserted in the cost function, in order to avoid a factor of 2 in its gradient.

[3]The term "independent" here means that these variables are (considered as) "functionally independent," in the sense that they evolve independently from one another: here, we are not talking about the statistical independence of random variables.

5.3   Methods with Constrained Source Variables

5.3.1   Methods with Nonnegativity (and Other) Constraints

Unlike the above methods, the approach considered here exploits the structure of $\tilde{S}$ as follows. The rows of $\tilde{S}$ and thus of $\check{S}$ may be seen as vectors used to decompose the row vectors of X, whereas $\tilde{A}$ and $\check{A}$ contain the coefficients of these decompositions. Moreover, matrix $\tilde{S}$ is guaranteed to be constrained: as shown by (2.12) and (2.19), only its top M rows are free, i.e., they contain the values of the original sources, whereas all subsequent rows are element-wise products of the above rows. Therefore, the same constraint is here set on the adaptive variable $\check{S}$ of the considered separating structure. This means that the top M rows of $\check{S}$ contain master, i.e., freely tuned, variables. These M row vectors are, respectively, denoted as $\check{s}_1$ to $\check{s}_M$. In contrast, all subsequent rows of $\check{S}$ contain slave variables, which are updated together with the above top M rows, so as to contain element-wise products (i.e., Hadamard products [92], denoted as $\odot$) $\check{s}_j \odot \check{s}_k$ of those top M rows. These $\check{s}_j \odot \check{s}_k$ products are stored only for $1 \le j \le k \le M$ for linear-quadratic mixtures, or $1 \le j < k \le M$ for bilinear mixtures. They are arranged in a fixed, arbitrarily selected, order (see, e.g., [82] for the natural order).

The free, i.e., master, variables to be optimized thus consist of:

- The top M rows of $\check{S}$. This sub-matrix of $\check{S}$ is hereafter denoted as $\check{S}_\ell$, since it aims at estimating "the linear part of $\tilde{S}$," i.e., the same matrix S as in the case of linear mixtures (see (2.8) and (2.15)).
- The complete matrix $\check{A}$.

These matrices may again be adapted so as to minimize $||X - \check{A}\check{S}||_F$ or its square. Moreover, in most reported investigations, this optimization was again performed subject to nonnegativity constraints imposed on all source and mixing coefficient variables. The separation criterion of the Linear-Quadratic Source-Constrained Nonnegative Matrix Factorization[4] (LQ-SC-NMF) method thus obtained reads

$$\left\{ \widehat{\check{A}}, \hat{S} \right\}_{LQ-SC-NMF} = \underset{\{\check{A}, \check{S}_\ell\}}{\operatorname{argmin}} \frac{1}{2} ||X - \check{A}\check{S}||_F^2 \qquad \text{subject to } \check{A} \ge 0, \check{S}_\ell \ge 0,$$

$$(5.3)$$

---

[4]It should be clear that, in all the terminology related to "Linear-Quadratic...Matrix Factorization" and "Bilinear...Matrix Factorization" introduced hereafter, the terms "Linear-Quadratic" and "Bilinear" refer to the properties of the observed data only with respect to the source values, not with respect to the overall set of quantities composed of these source values and of the mixing parameters. Also considering these mixing parameters, one might choose another terminology, because, e.g., mixtures which are linear with respect to the source values only are also bilinear with respect to the overall set of quantities composed of the source values and mixing parameters. This topic is discussed in [32], where the terminology "Linear-Quadratic *Mixture* Matrix Factorization" (LQMMF) and "Bilinear *Mixture* Matrix Factorization" (BMMF) is therefore used to avoid any ambiguity.

where $\widehat{\widehat{S}}$ and $\widecheck{S}$ are, respectively, derived from $\hat{S}$ and $\widecheck{S}_\ell$ by assigning their bottom rows to the element-wise products of their top rows, as explained above. This includes the case of bilinear mixtures, which leads to Bilinear Source-Constrained Nonnegative Matrix Factorization (B-SC-NMF) methods.

As suggested in Sect. 5.2, the criterion (5.3) obtained here may at first sight appear to be essentially the same as the criterion (5.2) defined in Sect. 5.2 for LE-NMF, but they have a difference which has major consequences for associated optimization algorithms: the cost function $||X - \widecheck{A}\widecheck{S}||_F^2$ to be optimized in (5.3) includes the matrix $\widecheck{S}$, whose bottom row elements depend on (more precisely, are products of) the elements of the variable $\widecheck{S}_\ell$ with respect to which this optimization is performed. The gradient of this cost function, which is used in various optimization algorithms, thus takes a different and significantly more complex form than in the LE-NMF approach, where the same cost function $||X - \widecheck{A}\widecheck{S}||_F^2$, optimized in (5.2), also includes the quantity $\widecheck{S}$ but is then optimized with respect to $\widecheck{S}$ itself. This added complexity of LQ-SC-NMF becomes even higher for optimization schemes such as the Newton algorithm, which involve the Hessian, i.e., *second-order* derivatives, associated with the above cost functions. The above comments concern derivatives with respect to source estimation parameters. In contrast, concerning the parameters used to estimate the mixing coefficients, the LE-NMF and LQ-SC-NMF methods have the same complexity, since their cost functions depend in the same way on the variable $\widecheck{A}$. As in the case of linear mixtures, various algorithms were derived from the above LQ-SC-NMF criterion (5.3), namely gradient-based, Newton, and multiplicative algorithms. They are especially detailed in [82]. That investigation of this type of algorithms moreover takes into account other constraints, which are imposed due to the considered remote sensing application:

1. In each observed signal, all linear coefficients sum to one:

$$\sum_{j=1}^{M} a_{ij} = 1 \qquad \forall\, i \in \{1, \ldots, P\}. \qquad (5.4)$$

2. All quadratic coefficients b_{ijk} have an upper bound:

$$b_{ijk} \leq 0.5 \quad \forall\, i \in \{1, \ldots, P\}, \quad \forall\, j \in \{1, \ldots, M\}, \quad \forall\, k \in \{j, \ldots, M\}. \qquad (5.5)$$

Besides, a similar approach is reported in [74], which deals with a very specific configuration involving two mixtures of two original sources, with the same quadratic contribution in both mixtures. Moreover, a modified version of this type of approach was proposed in [62] for uncorrelated sources. To this end, the above LQ-SC-NMF criterion (5.3) is modified by adding, to its cost function, regularization terms based on the correlation coefficients of pairs of estimated sources.

5.3.2 Methods Without Nonnegativity Constraints

Linear-quadratic matrix factorization methods, including bilinear ones, were initially developed as nonlinear extensions of linear NMF methods, thus including the above-mentioned nonnegativity constraints. However, they were later reconsidered without such constraints (and without (5.4), (5.5)), in the case of bilinear mixtures [30, 32] and linear-quadratic ones [32]. The separation principle then still consists in adapting $\hat{A}$ and the linear part $\check{S}_\ell$ of $\check{S}$ so that the product $\check{A}\check{S}$ fits the observed data matrix X, taking into account the structure of $\tilde{S}$ as in Sect. 5.3.1. The separation criterion of the standard Bilinear Source-Constrained Matrix Factorization (B-SC-MF) method thus obtained reads

$$\left\{\widehat{\tilde{A}}, \hat{S}\right\}_{B-SC-MF} = \operatorname*{argmin}_{\{\check{A}, \check{S}_\ell\}} \frac{1}{2}||X - \check{A}\check{S}||_F^2, \tag{5.6}$$

where, again, $\widehat{\tilde{S}}$ and $\check{S}$ are, respectively, derived from $\hat{S}$ and $\check{S}_\ell$ by assigning their bottom rows to the element-wise products of their top rows.

This deserves the following comments, which concern the considered Bilinear Matrix Factorization (BMF) separation principle and therefore the complete class of methods that may be derived from it, by using different associated separation criteria (not only criterion (5.6)) and algorithms. For linear mixtures (of mutually unconstrained sources), we showed in Sect. 5.2 that the version of the Matrix Factorization principle without nonnegativity constraints is not relevant, because it yields very high indeterminacies. In contrast, one may expect bilinear mixtures to possibly yield different properties, precisely due to the structure of the considered data, that is of the source matrix $\tilde{S}$ and therefore of the associated tunable matrix $\check{S}$. This may first be intuitively explained as follows. For an arbitrary value of the top M rows of $\check{S}$, the matrix product $\check{A}\check{S}$ yields row vectors which are linear combinations of the M vectors $\check{s}_1$ to $\check{s}_M$ and of their element-wise products $\check{s}_j \odot \check{s}_k$. Let us, e.g., consider the undesired case when each vector $\check{s}_j$ is not collinear to one of the actual source vectors which compose the top M rows of $\tilde{S}$, but is a linear combination of the latter vectors. Then, one may hope that the following property is met: the vector products $\check{s}_j \odot \check{s}_k$ have a "complex form" and are thus outside the subspace spanned by the actual source vectors and their element-wise products, i.e., outside the subspace spanned by the rows of X. Then, the product $\check{A}\check{S}$ may not exactly fit the observation matrix X, whatever the value of $\check{A}$ (unlike in the linear case of Sect. 5.2). Therefore, conversely, the exact fit $\check{A}\check{S} = X$ may be hoped to be achieved only when $\check{S}$ extracts the source signals, up to scaling and permutation.

Beyond the above intuitive approach, this problem was first analytically studied in [30], for the specific configuration consisting of bilinear mixtures of $M = 2$ sources. The following property was thus proved: if the eight vectors in the set $\{s_1, s_2, s_1 \odot s_1, s_1 \odot s_2, s_2 \odot s_2, s_1 \odot s_1 \odot s_2, s_1 \odot s_2 \odot s_2, s_1 \odot s_1 \odot s_2 \odot s_2\}$ are linearly independent and if the row vectors of X actually span the subspace defined by the three vectors $\{s_1, s_2, s_1 \odot s_2\}$,[5] then the condition set by the BMF separation principle, i.e., $\check{A}\check{S} = X$ (with the above-defined structure of $\check{S}$), is met if and only if the considered separating structure restores the sources up to scale factors and permutation. In other words, this BMF separation principle then ensures separability, i.e., uniqueness of the BMF decomposition up to the above indeterminacies. This result was then extended to an arbitrary number of sources in [32]. We again stress that, unlike in the linear case, this separability is obtained without setting such constraints as nonnegativity of the data. Similarly, this approach does not impose the properties which are used to build other classes of BSS methods, namely source statistical independence or sparsity. The above-defined structure of $\tilde{S}$ and $\check{S}$ is therefore of major importance: it is the reason why this separating structure for bilinear mixtures does not suffer from the unacceptably high indeterminacies that we highlighted in Sect. 5.2 for linear mixtures. Therefore, although the nonlinearity of the mixing model is traditionally considered as an issue, coming in addition to the usual difficulties already faced when dealing with *linear* BSS, the above discussion yields an original outcome from this point of view: it shows that mixture nonlinearity can be *exploited* in some cases in order to introduce new BSS approaches that are not applicable to linear mixtures.

The above discussion concerns bilinear mixtures. In contrast, [32] shows that the above approach yields spurious solutions for linear-quadratic mixtures.

As stated above, for bilinear mixtures of two sources, the method described in this section is guaranteed to achieve BSS if the eight vectors in the set $\{s_1, s_2, s_1 \odot s_1, s_1 \odot s_2, s_2 \odot s_2, s_1 \odot s_1 \odot s_2, s_1 \odot s_2 \odot s_2, s_1 \odot s_1 \odot s_2 \odot s_2\}$ are linearly independent.[6] It should be noted that this unusual and attractive property is also achieved by a quite different method, which was first reported for two sources in [51]. The latter method was also developed with a deterministic framework, but it uses the separating system of Sect. 3.1, with three mixtures of the source vectors s_1 and s_2, instead of the "direct model" of NMF considered up to now in the present section. The separation principle used in the method of [51] consists in tuning the coefficients of the separating system so as to reach the global maximum of the squared cross-correlation coefficient defined between (1) the product of the first two outputs of the separating system, which, respectively, estimate s_1 and s_2 or vice versa and (2) the third output of the separating system, which estimates $s_1 \odot s_2$. This BSS method and the analysis of its separability properties were then extended to an arbitrary number of sources in [35].

---

[5]This only requires X to contain at least three, linearly independent, row vectors.

[6]And if the row vectors of X actually span the subspace defined by the three vectors $\{s_1, s_2, s_1 \odot s_2\}$.

5.4   Methods with Constrained Source and Mixture Variables

5.4.1   Methods Without Nonnegativity Constraints

Still considering approaches without nonnegativity constraints, other Matrix Factorization methods were proposed in [30] and then exploited, e.g., in [32], by further constraining the structure of the separating system, i.e., its set of slave variables. More precisely, in the methods described in Sect. 5.3, both $\check{A}$ and the top M rows of $\check{S}$ are master, i.e., independently updated, variables. However, since this adaptation aims at minimizing the cost function

$$J_1 = \frac{1}{2}||X - \check{A}\check{S}||_F^2, \tag{5.7}$$

a different adaptation scheme may be used. In this new scheme, only the top M rows of $\check{S}$ are considered as master variables, whereas $\check{A}$ also becomes a slave variable. In each step of the loop for updating $\check{S}$, the slave variable $\check{A}$ is set to its optimum value, i.e., to its value which minimizes $||X - \check{A}\check{S}||_F$ with respect to $\check{A}$ for the considered value of $\check{S}$. This optimum is nothing but the least squares (LS) solution, i.e., (assuming $\check{S}$ has full row rank) [92]

$$\widehat{\check{A}}_{LS} = X\check{S}^T (\check{S}\check{S}^T)^{-1}. \tag{5.8}$$

Setting $\check{A} = \widehat{\check{A}}_{LS}$ in (5.7), the cost function to be optimized (only with respect to the top M rows of $\check{S}$) becomes

$$J_2 = \frac{1}{2}||X\left(I - \check{S}^T\left(\check{S}\check{S}^T\right)^{-1}\check{S}\right)||_F^2. \tag{5.9}$$

In this Bilinear Source-and-Mixture-Constrained Matrix Factorization (B-SMC-MF) approach, the separation criterion thus reads

$$\{\hat{S}\}_{B-SMC-MF} = \underset{\check{S}_\ell}{\mathrm{argmin}} \frac{1}{2}||X\left(I - \check{S}^T\left(\check{S}\check{S}^T\right)^{-1}\check{S}\right)||_F^2. \tag{5.10}$$

In this approach, $\widehat{\check{S}}$ and $\check{S}$ are again, respectively, derived from $\hat{S}$ and $\check{S}_\ell$ by assigning their bottom rows to the element-wise products of their top rows. Moreover, if one is also interested in BMI, i.e., in the estimation of the mixing coefficients, these coefficients are derived from the final value of $\check{S}$ (i.e., $\widehat{\check{S}}$) by means of (5.8).

This approach is attractive, first because the number of master variables adapted when using $\check{A} = \widehat{\check{A}}_{LS}$ and therefore J_2 is much lower than when using J_1: the set of tunable parameters here only consists of $\check{S}_\ell$, instead of $\check{S}_\ell$ and $\check{A}$ in (5.6). The searched space thus has a much lower dimension, which may decrease computational time and improve convergence properties, although this also depends on the shape of the variations of the considered cost function. Moreover, $\breve{A}$ and J_2 are thus defined by a closed-form expression, which allows one to derive the gradient of J_2 with respect to the master part $\breve{S}_\ell$ of $\breve{S}$. This gradient may then be used in gradient-based optimization algorithms, as was done in the modified configuration presented below in Sect. 5.4.2. Instead, the algorithm used in [30, 32] for the separation criterion (5.10) considered here is a derivative-free method, namely the Nelder-Mead algorithm.

5.4.2 Methods with Nonnegativity Constraints

The above approach was then modified in [13]. First, a nonnegativity constraint on the master source variables in $\breve{S}_\ell$ was introduced, thus leading to a different separation criterion, which reads

$$\{\hat{S}\}_{B-SMC-NSMF} = \underset{\breve{S}_\ell}{\operatorname{argmin}} \frac{1}{2}||X(I - \breve{S}^T(\breve{S}\breve{S}^T)^{-1}\breve{S})||_F^2 \quad \text{subject to } \breve{S}_\ell \geq 0.$$

(5.11)

This criterion thus achieves Bilinear Source-and-Mixture-Constrained Nonnegative-Source Matrix Factorization (B-SMC-NSMF). More precisely, the above criterion (5.11) imposes a nonnegativity constraint only on the source variables (on their master part and therefore on their products, i.e., slave variables). In contrast, it does not explicitly enforce such a constraint on the variables corresponding to the mixing coefficients: instead, these variables are implicitly set to (5.8), since the considered cost function is (5.9). Besides, in [13], a gradient-based algorithm was proposed to optimize the above cost function. A similar approach was proposed in [14] for linear-quadratic mixtures.

Moreover, [45] describes an NMF-based method which also involves constraints on both the source and mixing coefficient variables, but these constraints are different from those considered above concerning the coefficients. More precisely, that investigation concerns the specific bilinear-bilinear mixing model defined in Sect. 2.1. The mixing coefficients are thus constrained to meet condition $b_{ijk} = a_{ij}a_{ik}$. Therefore, the same constraint is imposed on the corresponding variables in the matrix factorization of the observations derived in that BSS method. All master variables are adapted by using a projected gradient algorithm. For the extended version of the bilinear-bilinear mixing model defined in Sect. 2.1, the separation method proposed in [91] especially combines the above-defined NMF principle and the concept of sparsity, which is discussed in Chap. 6.

Other approaches based on matrix factorization may, e.g., be found in [93].

Chapter 6
Sparse Component Analysis Methods

For linear mixtures, two major principles used in the literature for performing Sparse Component Analysis (SCA) may be briefly defined as follows. The first one consists in minimizing a sparsity-based cost function, such as the L0 pseudo-norm of an "error term." The second one takes advantage of zones (i.e., adjacent samples) in the sources where only one source is active, i.e., non-zero. These two principles have been extended to linear-quadratic mixtures and the resulting methods are presented hereafter. Besides, [65] deals with an approach which also takes advantage of small parts of the observed data where only one "contribution" is non-zero, more precisely pixels which correspond to only one pure material (i.e., pure pixels) in the considered application to unmixing of remote sensing spectra. However, the proposed criterion is only guaranteed to yield a *necessary* condition for detecting pure pixels.

6.1 A Method Based on L0 pseudo-Norm

The BSS method proposed in [44] is presented for a bilinear mixture of two original sources $s_1(n)$ and $s_2(n)$, involving three observations, so that the mixture is determined with respect to the extended set of sources. This makes it possible to use a *linear* separating structure, which is closely connected to the structure defined in Sect. 3.1. More precisely, the overall proposed approach consists of the following two stages.

First, linear combinations $z_{ij}(n)$ of two observed signals $x_i(n)$ and $x_j(n)$ are computed as

$$z_{ij}(n) = x_i(n) - c_{ij}x_j(n). \tag{6.1}$$

© The Author(s), under exclusive license to Springer Nature Switzerland AG 2021
Y. Deville et al., *Nonlinear Blind Source Separation and Blind Mixture Identification*, SpringerBriefs in Electrical and Computer Engineering, https://doi.org/10.1007/978-3-030-64977-7_6

A transform is then applied to them, which yields signals z'_{ij}. More precisely, a time-to-frequency transform is used. Besides, the transformed versions s'_1 and s'_2 of the original source signals are assumed to meet some sparsity conditions. Generally speaking, the type of sparsity considered in this investigation refers to the fact that at least some of the components of s'_1 and s'_2 should be equal to zero. More precise (sufficient) sparsity conditions are derived in [44]. Whenever z_{ij} still contains a term proportional to the product of two original source signals, this yields a term related to the convolution of s'_1 and s'_2 in z'_{ij}. This convolved term results in a more spread spectrum, i.e., it decreases the sparsity of z'_{ij}. Therefore, to remove the second-order term in z_{ij}, this suggests one to adapt the coefficients c_{ij} used to derive z_{ij} from x_i and x_j so as to maximize the sparsity of z'_{ij}. In the theoretical part of [44], the sparsity of a signal is measured by its L0 pseudo-norm, i.e., its number of non-zero components. The criterion used to select each coefficient c_{ij} is thus

$$\hat{c}_{ij} = \operatorname*{argmin}_{c_{ij}} ||z'_{ij}||_0. \tag{6.2}$$

This minimization is performed by means of an exhaustive search with respect to c_{ij} in [44]. In practice, an approximation of the L0 pseudo-norm is used in the above criterion, instead of the L0 pseudo-norm itself, since sparse signals have components that are close to zero, but not strictly equal to zero.

For each pair of observed signals x_i and x_j, the above first stage yields a signal z_{ij} which is restricted to a *linear* mixture of the source signals. The second stage of the proposed method then consists in applying a *linear* BSS method (depending on the considered source properties) to all signals z_{ij}, in order to separate the linearly mixed sources.

6.2 Methods Based on Single-Source Zones

As mentioned above, in the framework of linear memoryless mixtures, various SCA methods based on single-source zones were proposed, using deterministic or probabilistic representations of the source signals. This especially includes the LI-TempROM and LI-TiFROM methods (see especially [2–5]), whose names refer to the fact that they are intended for Linear Instantaneous mixtures and based on Temporal or Time-Frequency Ratios of Mixtures. This class of methods also contains LI-TempCORR and LI-TiFCORR [27, 36], which are applicable to Linear Instantaneous mixtures and based on Temporal or Time-Frequency Correlation-based parameters. A broader survey of this class of SCA methods is available in [29].

The above class, e.g., extends to the bilinear (memoryless) mixing model which is determined with respect to the extended set of sources. In particular, [33] describes versions of these methods expressed for stochastic source signals and operating in the original signal representation domain, which is hereafter called the "time domain," to define terminology more easily. Any sample of a signal thus corresponds to a given "time n." Moreover, as the methods intended for linear mixtures, their extensions considered here operate with parts of the signals corresponding to "analysis zones." Here, each temporal analysis zone is restricted to a single time n in the considered theoretical statistical framework. However, in practice, all signal moments are estimated over time intervals, and each temporal analysis zone then consists of such an interval. The methods considered here are based on correlation parameters, as the LI-TempCORR method that they extend. Unlike various methods described in this book, these methods are not restricted to the minimization of a single cost function: they consist of successive steps and they require a more in-depth description, which is therefore provided in Appendix A for the sake of readability.

Chapter 7
Extensions and Conclusion

In addition to the methods described in the previous chapters, other investigations reported in the literature deal with extensions of the mixing models of Chap. 2 studied so far in this book. Still considering second-order memoryless mixing models, an extension concerns the case when the source signals have variability (called spectral or intraclass variability by the remote sensing community), i.e., each observed signal is a combination of a version of the source signals which (somewhat) depends on the considered observed signal. An extension of NMF-based linear-quadratic BSS methods aiming at handling this variability was reported in [90] and the relationships between nonlinear and linear models with variability were studied in [38]. Besides, higher-order polynomial mixing models have been addressed in a few reported investigations, e.g., to handle multiple (beyond second-order) reflections in spectral unmixing applications: see [75] and the references therein for the use of a specific polynomial model and of its extension based on Fourier series; see also [10] and its extension [96] for non-blind configurations, and the references therein. In particular, several above-defined principles extend relatively straightforwardly to general polynomial mixing models. The current main trends in this emerging field of polynomial BSS may be summarized as follows:

- Most of the separating structures defined in Chap. 3 extend to polynomial mixtures, at the expense of increased complexity and/or by requiring an increased number of observations for the mixture to be determined with respect to the large number of extended sources (i.e., multi-source monomial terms) then faced. This includes:

  - extended versions of the linear separating system of Sect. 3.1 (which were not yet used in the literature for general polynomial models, to our knowledge),
  - the numerical inversion methods mentioned in Sect. 3.3, which are general-purpose and therefore not restricted to linear-quadratic mixtures (these methods have also presumably not yet been used in the literature for general polynomial mixtures),

© The Author(s), under exclusive license to Springer Nature Switzerland AG 2021 53
Y. Deville et al., *Nonlinear Blind Source Separation and Blind Mixture Identification*, SpringerBriefs in Electrical and Computer Engineering, https://doi.org/10.1007/978-3-030-64977-7_7

- dedicated nonlinear recurrent networks: as stated in Sect. 3.4 and detailed in [34], tailored versions of these networks may be developed for much wider classes of mixing models than linear-quadratic ones, including general polynomial models,
- factorization-based "separating structures" used for fitting the considered mixing model: an illustration of this approach was provided in [80] for a third-order polynomial model.

Different types of adaptation methods may then be developed to tune the parameters of the above separating structures. The methods reported so far in the literature cover as follows the classes of BSS methods described above in Chaps. 4–6.

- The first aspect concerns statistical methods. Reference [17] first investigates the general problem of inverting a multiple-input multiple-output memoryless polynomial mixing model. It then proceeds to BSS, by considering mutually statistically independent and binary-valued sources, focusing on the case of bilinear mixtures of two original sources. Also considering mutually statistically independent sources, an approach based on mutual information minimization is also proposed for a class of polynomial mixtures in [97]. Still using mutual information minimization, [9] extends the above-mentioned approach of [8] to specific polynomials, which are especially faced when considering ions with different valences in ion-selective electrode measurements.
- After introducing the above-mentioned factorization-based "separating structure" for polynomial mixtures, [80] also outlines an NMF-based procedure for adapting the parameters of that structure, again so as to fit the considered mixing model.
- As discussed in [29], SCA methods may be extended to different types of nonlinear mixtures, including polynomial ones, by using the principles that we described in Chap. 6 and Appendix A for linear-quadratic mixtures. The practical applicability of the SCA methods thus derived for higher-order mixtures may, however, be limited by the extended sparsity properties that they require. For instance, already for second-order mixtures, the approach proposed in subsection "A method based on non-stationarity conditions" to perform the "Remaining BMI and BSS tasks" of Sect. A.3 requires times when only two sources are simultaneously active, in addition to the request for times when only one source is active which is faced for linear mixtures with this type of methods.
- Other approaches may be found, e.g., in [10, 55] for non-blind configurations.

Finally, another extension beyond linear-quadratic memoryless mixtures was, e.g., reported in [84]. It concerns mixtures which are also restricted to second-order polynomials in terms on nonlinearity, but which are not "memoryless." More precisely, the considered source signals are two-dimensional, i.e., images. Each observed mixed value corresponds to a given "two-dimensional sample index," i.e., to a pixel position. It does not depend only on the source values for this sample index, but also on the source values for adjacent samples indices, i.e., adjacent pixels. This corresponds to the blurring effect, and the proposed separating system therefore includes associated processing means for achieving deblurring, in addition to the means intended for compensating for the nonlinearity of the mixture, modelled by second-order polynomials.

The topic of linear-quadratic BSS and BMI methods has therefore already been studied in detail, and this is the first step toward different types of extensions. Moreover, linear-quadratic mixtures appear in various applications, which were a strong motivation for investigating such mixtures, despite their increased complexity as compared with linear mixtures. As mentioned in Chap. 1, these applications especially concern remote sensing and the unsupervised unmixing of hyperspectral (and multispectral) images until now. One may anticipate that the considered types of mixing models and associated applications will extend in the future, as the general field of nonlinear signal/data processing keeps on growing.

Appendix A
Bilinear Sparse Component Analysis Methods Based on Single-Source Zones

We hereafter detail the SCA methods that we introduced in Sect. 6.2. We first define the signals that they involve, the assumptions used in these methods and therefore the associated terminology for sparsity concepts. We then explain how the different versions of this type of methods operate.

A.1 Considered Signals

The observations are here defined in scalar form by (2.3), or equivalently in matrix form by (2.11). Besides, we consider the centered version of these observations, defined as

$$x_i'(n) = x_i(n) - E\{x_i(n)\} \qquad \forall i \in \{1, \ldots, P\}. \tag{A.1}$$

Equations (2.3) and (A.1) then yield

$$x_i'(n) = \sum_{j=1}^{M} a_{ij} s_j'(n) + \sum_{j=1}^{M-1} \sum_{k=j+1}^{M} b_{ijk} p_{jk}'(n) \qquad \forall i \in \{1, \ldots, P\}, \tag{A.2}$$

where $s_j'(n)$ and $p_{jk}'(n)$ are, respectively, the centered versions of $s_j(n)$ and $p_{jk}(n) = s_j(n) s_k(n)$. This yields in matrix form

$$x'(n) = As'(n) + Bp'(n), \tag{A.3}$$

where the vectors $x'(n)$, $s'(n)$ and $p'(n)$ are the centered versions of those involved in (2.11).

© The Author(s), under exclusive license to Springer Nature Switzerland AG 2021
Y. Deville et al., *Nonlinear Blind Source Separation and Blind Mixture Identification*, SpringerBriefs in Electrical and Computer Engineering, https://doi.org/10.1007/978-3-030-64977-7

A.2   Definitions and Assumptions, Sparsity Concepts

Assumption A.1 *All source signals $s_1(n), \ldots, s_M(n)$ are zero mean at any time n.*[1]

Definition A.1 A signal is "active" at time n if it has non-zero mean power at that time. It is "inactive" at time n if it has zero mean power at that time.

Definition A.2 A source signal $s_j(n)$ is "isolated" in an analysis zone if only this source signal (among all the signals $s_1(n), \ldots, s_M(n)$) is active in this analysis zone. An analysis zone where a source signal is isolated, i.e., where a single source signal is active, is called a "single-source zone." An analysis zone where several source signals are active is called a "multiple-source zone."

The above definition of an isolated source signal corresponds to the theoretical point of view. From a practical point of view, this means that the mean powers of all other source signals are negligible as compared with the mean power of the source signal that is isolated.

Definition A.3 A source signal $s_j(n)$ is "accessible" in the temporal representation domain if there exists at least one analysis zone inside this domain where this source signal is isolated.

Assumption A.2 *All source signals $s_1(n), \ldots, s_M(n)$ are accessible in the temporal representation domain.*

In other words, the only sparsity constraint set at this stage by these methods is that, for each source, there should exist a tiny zone, in the temporal representation domain, where only this source is active. In all other zones, these methods allow several sources to overlap, i.e., to be simultaneously active. The constraint on joint sparsity thus set on the sources is very low and this type of "sparse component analysis" methods might therefore be more precisely called "quasi-non-sparse component analysis methods." Besides, the considered sources are thus requested to be non-stationary,[2] since their mean powers are zero at some times and non-zero at others. If all sources have zero mean power in a given analysis zone, all observations also have zero mean power in that zone and some parameters used in the considered SCA methods are then undefined. To avoid this situation, the following "technical assumption" is made:

Assumption A.3 *On each analysis zone, at least one source is active.*

---

[1]The observations may then be non-zero mean, due to the nonlinear nature of the mixing model and the possible source correlation. This is a motivation for using the centered version $x_i'(n)$ of the observations.

[2]More precisely, they are long-term non-stationary, but they should be short-term stationary in practice, in order to make it possible to estimate the required signal moments, defined below, over short time intervals.

Finally, the extended source signals should have some "diversity" as defined hereafter:

Assumption A.4 *For any considered time n, the signals which are contained in $s'(n)$ and $p'(n)$ and which are active at that time are linearly independent (if there exist at least two such active signals at that time).*

This assumption is based on the definition of linear independence of random variables, e.g., provided in [29]. It means that these SCA methods also apply to situations where the active signals in $s'(n)$ and $p'(n)$ are correlated, which is an attractive feature as compared with ICA methods.

These SCA methods also use the following assumptions, concerning the mixing matrix: all mixing coefficients a_{ij} are non-zero and A is a full-column-rank matrix.

A.3   SCA Methods

A.3.1   Identification of Linear Part of Mixture

The first step of the considered methods consists in identifying the "linear part" of the mixing model, i.e., the matrix A, or more precisely the matrix $C = [c_{ij}]$, where

$$c_{ij} = \frac{a_{i,\sigma(j)}}{a_{1,\sigma(j)}} \quad \forall i \in \{1, \ldots, P\}, \quad \forall j \in \{1, \ldots, M\} \tag{A.4}$$

and $\sigma(.)$ is a permutation. C is therefore a modified version of A, where the columns are permuted and each column is rescaled with respect to the value in its first row, i.e., with respect to its linear contribution in observation $x_1(n)$.

As shown in [33], despite the presence of the quadratic part of the mixing model, the linear part C of this model may be identified by the same type of procedure as in the above-mentioned LI-TempCorr method intended for linear memoryless mixtures (here using the centered version of the signals). This procedure for linear mixtures, which is detailed in [33], is therefore skipped in this book, where we directly proceed to the aspects of the BMI and BSS tasks which are specific to the bilinear memoryless mixing model.

A.3.2   Cancellation of Linear Part of Mixture

The considered SCA methods then derive a set of L signals $z_\ell(n)$ from the observations $x_i(n)$, in such a way that these signals $z_\ell(n)$ only contain quadratic cross-terms, i.e., terms proportional to $p_{jk}(n)$. To this end, one considers signals defined as

$$z_\ell(n) = x_1(n) - \sum_{i=2}^{P} d_{\ell i} x_i(n) \quad \forall\, \ell \in \{1, \ldots, L\}. \tag{A.5}$$

Combining this expression with (2.3) and (A.4) yields

$$z_\ell(n) = \sum_{j=1}^{M} a_{1,\sigma(j)} s_{\sigma(j)}(n) \left[1 - \sum_{i=2}^{P} c_{ij} d_{\ell i} \right] + \sum_{j=1}^{M-1} \sum_{k=j+1}^{M} r_{\ell jk} p_{jk}(n)$$

$$\forall\, \ell \in \{1, \ldots, L\}. \tag{A.6}$$

To obtain a signal $z_\ell(n)$ which contains no linear terms associated with any $s_j(n)$, the coefficients $d_{\ell i}$ are selected so that

$$\sum_{i=2}^{P} c_{ij} d_{\ell i} = 1 \quad \forall\, j \in \{1, \ldots, M\}. \tag{A.7}$$

For a given index ℓ, this yields a set of M equations, where the unknowns are the $P - 1$ values of $d_{\ell i}$, whereas the (estimated) coefficients c_{ij} are available from the first step of these methods, i.e., from the identification of the linear part of the mixing model. If $P - 1 = M$, this set of linear equations has a single solution, i.e., one can only create one such signal $z_\ell(n)$. More generally speaking, whatever $M' \geq 0$, if $P - 1 = M + M'$, one can create $M' + 1$ linearly independent signals $z_\ell(n)$. Besides, (A.6) then reduces to

$$z_\ell(n) = \sum_{j=1}^{M-1} \sum_{k=j+1}^{M} r_{\ell jk} p_{jk}(n) \quad \forall\, \ell \in \{1, \ldots, L\}, \tag{A.8}$$

i.e., these signals $z_\ell(n)$ are then only mixtures of the quadratic signals $p_{jk}(n)$. Moreover, there exist $M(M-1)/2$ signals[3] $p_{jk}(n)$ in the observations (2.3). The set of mixtures $z_\ell(n)$ of the signals $p_{jk}(n)$ is requested to be invertible. To this end, the numbers L and P of recombined signals $z_\ell(n)$ and observations $x_i(n)$ are set to $L = M' + 1 = M(M-1)/2$ and therefore $P = M + M' + 1 = M(M+1)/2$.

The following result is thus obtained: by solving Eq. (A.7) and deriving the resulting signals according to (A.5), one obtains the set of linear memoryless mixtures $z_\ell(n)$ of the signals $p_{jk}(n)$ defined by (A.8), which is invertible when $[r_{\ell jk}]$ is assumed to be invertible. These mixed signals may then be used in various ways, as will now be shown.

---

[3] Or less if all coefficients for at least one signal $p_{jk}(n)$ are zero.

A.3.3   Remaining BMI and BSS Tasks

One may then proceed in different ways, depending on which parts of the BMI and BSS tasks should be performed in the considered application and which constraints on the sources are acceptable. Three alternative methods are described hereafter.

A.3.3.1   A Method Based on Non-stationarity Conditions

We first again focus on methods for signals which are time-domain sparse, and therefore non-stationary. One may then process the *linear* (memoryless) mixtures $z_\ell(n)$ of the signals $p_{jk}(n)$, defined in (A.8), by adapting the first step of the approach defined above in Sect. A.3.1, i.e., the step intended for the "Identification of linear part of mixture," to this new context. This achieves both BMI for the mixing matrix in (A.8) (but not yet for the original matrix B in (2.11)) and BSS for the signals $p_{jk}(n)$ (but not yet for the signals $s_j(n)$). This adaptation of the first step of the approach defined in Sect. A.3.1 requires one to extend the assumptions accordingly. Especially, one then needs times when a single signal $p_{jk}(n)$ is active, i.e., essentially times when only the two corresponding sources $s_j(n)$ and $s_k(n)$ are simultaneously active.

It should also be noted that in the basic configuration with $M = 2$ sources, only a single signal $p_{jk}(n)$ exists, namely $s_1(n)s_2(n)$. This signal is then directly provided by the method described above, so that the stage described in the current section then disappears.

A.3.3.2   A Method Also Using Other Correlation Parameters

The first method, defined in Sect. A.3.3.1 above, yields scaled permuted versions of the signals $p_{jk}(n)$, i.e., it provides a set of signals

$$y_\ell(n) = \lambda_{jk} p_{jk}(n) \quad \forall \ell \in \{1, \ldots, L\}. \tag{A.9}$$

The following simple method may then be applied to these signals when one also wants to identify the matrix B and/or to separate the signals $s_j(n)$. Considering the signals which are contained in $s'(n)$ and $p'(n)$ at times when they are active, they are here requested to be uncorrelated, unlike in the previous stages of this approach. Denoting $y'_\ell(n)$ the centered version of $y_\ell(n)$, we then have if $p_{jk}(n)$ is active

$$\delta_{i\ell} = \frac{E\{y'_\ell(n)x'_i(n)\}}{E\{[y'_\ell(n)]^2\}} = \frac{b_{ijk}}{\lambda_{jk}} \quad \forall i \in \{1, \ldots, P\}, \quad \forall \ell \in \{1, \ldots, L\}. \tag{A.10}$$

This may be interpreted as in the first step of the approach defined above in Sect. A.3.1, i.e., in the step intended for the "Identification of linear part of mixture":

one may build the matrix $[\delta_{i\ell}]$, where each column ℓ corresponds to one signal $p_{jk}(n)$. Equation (A.10) then shows that this matrix is equal to B, up to the scale and permutation indeterminacies. This completes all BMI tasks for the considered mixing model. Moreover, let us consider the signals

$$u_i(n) = x_i(n) - \sum_{\ell=1}^{L} \delta_{i\ell} y_\ell(n) \quad \forall\, i \in \{1, \ldots, P\}. \tag{A.11}$$

Denoting $u(n)$ the column vector of signals $u_i(n)$, Eqs. (2.3), (A.9), (A.10), and (A.11) then yield in matrix form

$$u(n) = As(n). \tag{A.12}$$

BSS is then straightforwardly achieved for the original sources $s_j(n)$ by computing the vector $C^\dagger u(n)$, where † denotes the pseudo-inverse.

A.3.3.3   A Method Only Using Variance Parameters

Eventually, if one is mainly interested in the BSS of the sources $s_j(n)$, the method of Sect. A.3.3.1 and its constraints may be avoided, again at the expense of requesting the uncorrelation of the signals which are contained in $s'(n)$ and $p'(n)$ (considered at times when they are active). To this end, one introduces the signals

$$v_i(n) = x_i(n) - \sum_{\ell=1}^{L} d_{i\ell} z_\ell(n) \quad \forall\, i \in \{1, \ldots, P\}. \tag{A.13}$$

It may be shown that, by adapting all coefficients $d_{i\ell}$ so as to minimize the variances of all signals $v_i(n)$, the vector $v(n)$ consisting of these signals becomes equal to $As(n)$. BSS is then achieved for the original sources $s_j(n)$ by computing the vector $C^\dagger v(n)$.

Bibliography

1. K. Abed-Meraim, A. Belouchrani Y. Hua, Blind identification of a linear-quadratic mixture of independent components based on joint diagonalization procedure, in *1996 IEEE International Conference on Acoustics, Speech, and Signal Processing Conference Proceedings*, vol. 5, Atlanta, GA (1996), pp. 2718–2721
2. F. Abrard, Y. Deville, Blind separation of dependent sources using the "TIme-Frequency Ratio Of Mixtures" approach, in *Proceedings of the 7th International Symposium on Signal Processing and its Applications (ISSPA 2003)*, IEEE Catalog Number 03EX714C, ISBN 0-7803-7947-0, Paris (2003)
3. F. Abrard, Y. Deville, A time-frequency blind signal separation method applicable to underdetermined mixtures of dependent sources. Signal Process. **85**(7), 1389–1403 (2005)
4. F. Abrard, Y. Deville, P. White, A new source separation approach based on time-frequency analysis for instantaneous mixtures, in *Proceedings of the 5th International Worshop on Electronics, Control, Modelling, Measurement and Signals (ECM2S'2001)*, Toulouse, (2001), pp. 259–267
5. F. Abrard, Y. Deville, P. White, From blind source separation to blind source cancellation in the underdetermined case: a new approach based on time-frequency analysis, in *Proceedings of the 3rd International Conference on Independent Component Analysis and Signal Separation (ICA'2001)*, San Diego (2001), pp. 734–739
6. M.S.C. Almeida, L.B. Almeida, Separating nonlinear image mixtures using a physical model trained with ICA, in *Proceedings of the 16th IEEE Signal Processing Society Workshop on Machine Learning for Signal Processing (MLSP 2006)*, Arlington, VA (2006), pp. 65–70
7. M.S.C. Almeida, L.B. Almeida, Nonlinear separation of show-through image mixtures using a physical model trained with ICA. Signal Process. **92**, 872–884 (2012)
8. R. Ando, L.T. Duarte, C. Jutten, R. Attux, A blind source separation method for chemical sensor arrays based on a second order mixing model, in *Proceedings of the 23rd European Signal Processing Conference (EUSIPCO 2015)*, Nice (2015), pp. 938–942
9. R. Ando, C. Jutten, B. Rivet, R. Attux, L.T. Duarte, Nonlinear blind source separation for chemical sensor arrays based on a polynomial representation, in *Proceedings of the 24th European Signal Processing Conference (EUSIPCO 2016)*, Budapest (2016), pp. 2146–2150
10. V. Andrejchenko, Z. Zahiri, R. Heylen, P. Scheunders, A spectral mixing model accounting for multiples reflections and shadow, in *Proceedings of the 2019 IEEE International Geoscience and Remote Sensing Symposium (IGARSS 2019)*, Yokohama (2019), pp. 286–289
11. G. Bedoya, Non-linear blind signal separation for chemical solid-state sensor arrays. Ph.D. Thesis, Universitat Politecnica de Catalunya, 2006

© The Author(s), under exclusive license to Springer Nature Switzerland AG 2021
Y. Deville et al., *Nonlinear Blind Source Separation and Blind Mixture Identification*, SpringerBriefs in Electrical and Computer Engineering, https://doi.org/10.1007/978-3-030-64977-7

12. A. Belouchrani, K. Abed-Meraim, J.-F. Cardoso, E. Moulines, A blind source separation technique using second-order statistics. IEEE Trans. Signal Process. **45**(2), 434–444 (1997)
13. F.Z. Benhalouche, Y. Deville, M.S. Karoui, A. Ouamri, Bilinear matrix factorization using a gradient method for hyperspectral endmember spectra extraction, in *Proceedings of the 2016 IEEE International Geoscience and Remote Sensing Symposium (IGARSS 2016)*, Beijing (2016), pp. 6565–6568
14. F.Z. Benhalouche, Y. Deville, M.S. Karoui, A. Ouamri, Hyperspectral endmember spectra extraction based on constrained linear-quadratic matrix factorization using a projected gradient method, in *Proceedings of the 2016 IEEE International Workshop on Machine Learning for Signal Processing (MLSP 2016)*, Salerno (2016), pp. 1–6
15. J.-F. Cardoso, Infomax and maximum likelihood for blind source separation. IEEE Signal Process. Lett. **4**(4), 112–114 (1997)
16. J.F. Cardoso, The three easy routes to independent component analysis: contrasts and geometry, in *Proceedings of the 3rd International Conference on Independent Component Analysis and Signal Separation (ICA'2001)*, San Diego (2001)
17. M. Castella, Inversion of polynomial systems and separation of nonlinear mixtures of finite-alphabet sources. IEEE Trans. Signal Process. **56**(8), 3905–3917 (2008)
18. C. Chaouchi, Y. Deville, S. Hosseini, Nonlinear source separation: a quadratic recurrent inversion structure, in *Proceedings of the 9th International Worshop on Electronics, Control, Modelling, Measurement and Signals (ECMS 2009)*, Arrasate-Mondragon (2009), pp. 91–98
19. C. Chaouchi, Y. Deville, S. Hosseini, Une structure récurrente pour la séparation de mélanges quadratiques, in *Proceedings of GRETSI 2009*, Dijon (2009)
20. C. Chaouchi, Y. Deville, S. Hosseini, Nonlinear source separation: a maximum likelihood approach for quadratic mixtures, in *Proceedings of the 30th International Workshop on Bayesian Inference and Maximum Entropy Methods in Science and Engineering (MaxEnt 2010)*, Chamonix (2010)
21. S. Chaouchi, Y. Deville, S. Hosseini, Cumulant-based estimation of quadratic mixture parameters for blind source separation, in *Proceedings of the 18th European Signal Processing Conference (EUSIPCO 2010)*, Aalborg (2010), pp. 1826–1830
22. A. Cichocki, R. Zdunek, A.H. Phan, S.-I. Amari, *Nonnegative Matrix and Tensor Factorizations: Applications to Exploratory Multi-Way Data Analysis and Blind Source Separation* (Wiley, Chichester, 2009)
23. P. Comon, Independent Component Analysis, a new concept?. Signal Process. **36**(3), 287–314 (1994)
24. P. Comon, C. Jutten (eds.), *Handbook of Blind Source Separation: Independent Component Analysis and Applications* (Academic, Oxford, 2010)
25. Y. Deville, Système pour estimer des signaux reçus sous forme de signaux mélangés. Brevet français, priorité 22.02.1995 FR 9502051. Demande de brevet européen EP 0 729 037 A1. Y. Deville, "System for estimating signals received in the form of mixed signals", United States Patent US005909646A (1995)
26. Y. Deville, Méthode de séparation de sources pour mélanges linéaires-quadratiques ("a source separation method for linear-quadratic mixtures", in French), Private Communication (2000)
27. Y. Deville, Temporal and time-frequency correlation-based blind source separation methods, in *Proceedings of the Fourth International Symposium on Independent Component Analysis and Blind Signal Separation (ICA2003)*, Nara (2003), pp. 1059–1064
28. Y. Deville, *TRAITEMENT DU SIGNAL: Signaux temporels et spatiotemporels - Analyse des signaux, théorie de l'information, traitement d'antenne, séparation aveugle de sources* (Ellipses Editions Marketing, Paris, 2011)
29. Y. Deville, Chapter 6. Sparse component analysis: a general framework for linear and nonlinear blind source separation and mixture identification, in *Blind Source Separation: Advances in Theory, Algorithms and Applications*, ed. by G. Naik, W. Wang (Springer, Berlin, 2014), pp. 151–196
30. Y. Deville, Matrix factorization for bilinear blind source separation: methods, separability and conditioning, in *Proceedings of the 23rd European Signal Processing Conference (EUSIPCO 2015)*, Nice (2015), pp. 1945–1949

31. Y. Deville, Blind source separation and blind mixture identification methods, in *Wiley Encyclopedia of Electrical and Electronics Engineering*, ed. by J. Webster (Wiley, Hoboken, 2016), pp. 1–33
32. Y. Deville, From separability/identifiability properties of bilinear and linear-quadratic mixture matrix factorization to factorization algorithms. Digital Signal Process. **87**, 21–33 (2019)
33. Y. Deville, S. Hosseini, Blind identification and separation methods for linear-quadratic mixtures and/or linearly independent non-stationary signals, in *Proceedings of the 9th International Symposium on Signal Processing and its Applications (ISSPA 2007)*, IEEE Catalog Number 07EX1580C, ISBN 1-4244-0779-6, Sharjah (2007)
34. Y. Deville, S. Hosseini, Recurrent networks for separating extractable-target nonlinear mixtures. Part I: non-blind configurations. Signal Process. **89**(4), 378–393 (2009)
35. Y. Deville, S. Hosseini, Blind source separation methods based on output nonlinear correlation for bilinear mixtures of an arbitrary number of possibly correlated signals, in *Proceedings of the Eleventh IEEE Sensor Array and Multichannel Signal Processing Workshop (SAM 2020)*, Hangzhou (2020)
36. Y. Deville, M. Puigt, Temporal and time-frequency correlation-based blind source separation methods. Part I: determined and underdetermined linear instantaneous mixtures. Signal Process. **87**(3), 374–407 (2007)
37. Y. Deville, S. Hosseini, A. Deville, Effect of indirect dependencies on maximum likelihood and information theoretic blind source separation for nonlinear mixtures. Signal Process. **91**(4), 793–800 (2011)
38. L. Drumetz, B. Ehsandoust, J. Chanussot, B. Rivet, M. Babaie-Zadeh, C. Jutten, Relationships between nonlinear and space-variant linear models in hyperspectral image unmixing. IEEE Signal Process. Lett. **24**(10), 1567–1571 (2017)
39. L.T. Duarte, Design of smart chemical sensor arrays: an approach based on source separation methods, Ph.D., Grenoble, 2009
40. L.T. Duarte, C. Jutten, S. Moussaoui, Bayesian source separation of linear-quadratic and linear mixtures through a MCMC method, in *Proceedings of IEEE MLSP*, Grenoble (2009)
41. L.T. Duarte, C. Jutten, S. Moussaoui, Séparation de sources dans le cas de mélanges linéaires-quadratiques et linéaires par une approche bayésienne, in *Proceedings of GRETSI 2009*, Dijon (2009)
42. L.T. Duarte, R. Suyama, R. Attux, Y. Deville, J.M.T. Romano, C. Jutten, Blind source separation of overdetermined linear-quadratic mixtures, in *Proceedings of the 9th International Conference on Latent Variable Analysis and Signal Separation (LVA/ICA 2010)*. Lecture Notes in Computer Science, vol. 6365 (Springer, St. Malo, 2010), pp. 263–270
43. L. T. Duarte, C. Jutten, S. Moussaoui, Bayesian source separation of linear and linear-quadratic mixtures using truncated priors. J. Signal Process. Syst. **65**(3), 311–323 (2011)
44. L.T. Duarte, R.A. Ando, R. Attux, Y. Deville, C. Jutten, Separation of sparse signals in overdetermined linear-quadratic mixtures, in *Proceedings of the 10th International Conference on Latent Variable Analysis and Signal Separation (LVA/ICA 2012)*. Lecture Notes in Computer Science, vol. 7191 (Springer, Tel-Aviv, 2012), pp. 239–246
45. O. Eches, M. Guillaume, A bilinear-bilinear nonnegative matrix factorization method for hyperspectral unmixing. IEEE Geosci. Remote Sensing Lett. **11**(4), 778–782 (2014)
46. W. Fan, B. Hu, J. Miller, M. Li, Comparative study between a new nonlinear model and common linear model for analysing laboratory simulated-forest hyperspectral data. Int. J. Remote Sensing **30**(11), 2951–2962 (2009)
47. D.G. Fantinato, R.A. Ando, A. Neves, L.T. Duarte, C. Jutten, R. Attux, A quadratic divergence-based independence measure applied to linear-quadratic mixtures, in *Proceedings of XXXIV Simposio Brasileiro de Telecomunicacoes, SBrT 2016*, Santarem (2016), pp. 279–283
48. D.G. Fantinato, L.T. Duarte, B. Rivet, B. Ehsandoust, R. Attux, C. Jutten, Gaussian processes for source separation in overdetermined bilinear mixtures, in *Proceedings of the 13th International Conference on Latent Variable Analysis and Signal Separation (LVA/ICA 2017)*. Lecture Notes in Computer Science, vol. 10169 (Springer International Publishing AG, Grenoble, 2017), pp. 300–309

49. M. Gaeta, J.-L. Lacoume, Source separation without a priori knowledge: the maximum likelihood solution, in *European Signal Processing Conference (EUSIPCO)* (1990), pp. 621–624

50. P. Georgiev, Blind source separation of bilinearly mixed signals, in *Proceedings of ICA 2001*, San Diego (2001), pp. 328–331

51. A. Guerrero, Y. Deville, S. Hosseini, A blind source separation method based on output nonlinear correlation for bilinear mixtures, in *Proceedings of the 14th International Conference on Latent Variable Analysis and Signal Separation (LVA/ICA 2018)*. Lecture Notes in Computer Science, vol. 10891 (Springer International Publishing AG, part of Springer Nature, Guildford 2018), pp. 183–192

52. R. Guidara, S. Hosseini, Y. Deville, Blind separation of nonstationary Markovian sources using an equivariant Newton-Raphson algorithm, IEEE Signal Process. Lett. **16**(5), 426–429 (2009). Comments: see also Non-stationary Markovian blind source separation. Solving estimating equations using an equivariant Newton-Raphson algorithm, online Technical Report (2008). http://www.ast.obs-mip.fr/users/ydeville/papers/rgshyd_sept2008_tech_report.pdf

53. R. Guidara, S. Hosseini, Y. Deville, Maximum likelihood blind image separation using non-symmetrical half-plane Markov random fields. IEEE Trans. Image Process. **18**(11), 2435–2450 (2009)

54. A. Halimi, Y. Altmann, N. Dobigeon, J.-Y. Tourneret, Nonlinear unmixing of hyperspectral images using a generalized bilinear model. IEEE Trans. Geosci. Remote Sensing **49**(11), 4153–4162 (2011)

55. A. Halimi, J.M. Bioucas-Dias, N. Dobigeon, G.S. Buller, S. McLaughlin, Fast hyperspectral unmixing in presence of nonlinearity or mismodeling effects. IEEE Trans. Comput. Imaging **3**(2), 146–159 (2017)

56. J. Hérault, B. Ans, Circuits neuronaux à synapses modifiables: décodage de messages composites par apprentissage non-supervisé. C.R. de l'Académie des Sciences de Paris, t. **299**(13), 525–528 (1984)

57. S. Hosseini, Y. Deville, Blind separation of linear-quadratic mixtures of real sources using a recurrent structure, in *Proceedings of the 7th International Work-conference on Artificial And Natural Neural Networks (IWANN 2003)*, ed. by J. Mira, J.R. Alvarez, vol. 2 (Springer, Mao, Menorca, 2003), pp. 241–248

58. S. Hosseini, Y. Deville, Blind maximum likelihood separation of a linear-quadratic mixture, in *Proceedings of the Fifth International Conference on Independent Component Analysis and Blind Signal Separation (ICA 2004)*. Lecture Notes in Computer Science, vol. 3195 (Springer, Granada, 2004), pp. 694–701. ISSN 0302-9743, ISBN 3-540-23056-4. Erratum: see also "Correction to "Blind maximum likelihood separation of a linear-quadratic mixture"", available on-line at http://arxiv.org/abs/1001.0863

59. S. Hosseini, Y. Deville, Recurrent networks for separating extractable-target nonlinear mixtures. Part II: blind configurations. Signal Process. **93**(4), 671–683 (2013)

60. S. Hosseini, Y. Deville, Blind separation of parametric nonlinear mixtures of possibly autocorrelated and non-stationary sources. IEEE Trans. Signal Process. **62**(24), 6521–6533 (2014)

61. S. Hosseini, C. Jutten, D.T. Pham, Markovian source separation. IEEE Trans. Signal Process. **51**(12), 3009–3019 (2003)

62. S. Hosseini, Y. Deville, L.T. Duarte, A. Selloum, Extending NMF to blindly separate linear-quadratic mixtures of uncorrelated sources, in *Proceedings of the 2016 IEEE International Workshop on Machine Learning for Signal Processing (MLSP 2016)*, Salerno (2016), pp. 1–6

63. P. Huard, R. Marion, Study of non-linear mixing in hyperspectral imagery - a first attempt in the laboratory, in *Proceedings of the Third Workshop on Hyperspectral Image and Signal Processing: Evolution in Remote Sensing (WHISPERS 2011)*, Lisbon (2011)

64. A. Hyvärinen, J. Karhunen, E. Oja, *Independent Component Analysis* (Wiley, New York, 2001)

65. L. Jarboui, S. Hosseini, Y. Deville, R. Guidara, A. Ben Hamida, A new unsupervised method for hyperspectral image unmixing using a linear-quadratic model, in *Proceedings of the First International Conference of Advanced Technologies for Signal and Image Processing (ATSIP 2014)*, Sousse (2014), pp. 423–428

66. L. Jarboui, S. Hosseini, R. Guidara, Y. Deville, A. Ben Hamida, A MAP-based NMF approach to hyperspectral image unmixing using a linear-quadratic mixture model, in *Proceedings of the 2016 IEEE International Conference on Acoustics, Speech, and Signal Processing (ICASSP 2016)*, Shanghai (2016), pp. 3356–3360

67. L. Jarboui, Y. Deville, S. Hosseini, R. Guidara, A. Ben Hamida, L.T. Duarte, A second-order blind source separation method for bilinear mixtures. Multidim. Syst. Signal Process. **29**(3), 1153–1172 (2018)

68. C. Jutten, J. Hérault, Blind separation of sources, part I: an adaptive algorithm based on neuromimetic architecture. Signal Process. **24**(1), 1–10 (1991)

69. M. Kendall, A. Stuart, *The Advanced Theory of Statistics*, vol. 1 (Charles Griffin, London & High Wycombe, 1977)

70. M. Krob, M. Benidir, Blind identification of a linear-quadratic model using higher-order statistics, in *Proceedings of ICASSP 1993*, vol. 4, Minneapolis (1993), pp. 440–443

71. M. Krob, M. Benidir, Blind identification of a linear-quadratic mixture: application to quadratic phase coupling estimation, in *Proceedings of the 1993 IEEE Signal Processing Workshop on Higher-Order Statistics*, South Lake Tahoe, CA (1993), pp. 351–355

72. D.D. Lee, H.S. Seung, Learning the parts of objects by non-negative matrix factorization. Nature **401**, 788–791 (1999)

73. D.D. Lee, H.S. Seung, Algorithms for non-negative matrix factorization. Adv. Neural Info. Proc. Syst. **13**, 556–562 (2001)

74. Q. Liu, W. Wang, Show-through removal for scanned images using non-linear NMF with adaptive smoothing, in *Proceedings of 2013 IEEE China Summit & International Conference on Signal and Information Processing (ChinaSIP)*, Beijing (2013), pp. 650–654

75. H. Lu, Y. Li, F. Chen, H. Zhou, C. Cui, X. Zhu, A generalization of p-linear mixing model by combination of two kinds of approximator in hyperspectral unmixing, in *Proceedings of the 2018 IEEE International Geoscience and Remote Sensing Symposium (IGARSS 2018)*, Valencia (2018), pp. 2705–2708

76. S. Madrolle, P. Grangeat, C. Jutten, A linear-quadratic model for the quantification of a mixture of two diluted gases with a single metal oxide sensor. Sensors **18**(6), paper no. 1785 (2018)

77. S. Madrolle, L. Duarte, P. Grangeat, C. Jutten, A Bayesian blind source separation method for a linear-quadratic model, in *Proceedings of the 26th European Signal Processing Conference (EUSIPCO 2018)*, Rome (2018), pp. 1242–1246

78. A. Mansour, C. Jutten, A direct solution for blind separation of sources. IEEE Trans. Signal Process. **44**(3), 746–748 (1996)

79. I. Meganem, P. Déliot, X. Briottet, Y. Deville, S. Hosseini, Physical modelling and non-linear unmixing method for urban hyperspectral images, in *Proceedings of the Third Workshop on Hyperspectral Image and Signal Processing: Evolution in Remote Sensing (WHISPERS 2011)*, Lisbon (2011)

80. I. Meganem, Y. Deville, S. Hosseini, P. Déliot, X. Briottet, L. T. Duarte, Linear-quadratic and polynomial non-negative matrix factorization; application to spectral unmixing, in *Proceedings of the 19th European Signal Processing Conference (EUSIPCO 2011)*, Barcelona (2011)

81. I. Meganem, P. Déliot, X. Briottet, Y. Deville, S. Hosseini, Linear-quadratic mixing model for reflectances in urban environments. IEEE Trans. Geosci. Remote Sensing **52**(1), 544–558 (2014)

82. I. Meganem, Y. Deville, S. Hosseini, P. Déliot, X. Briottet, Linear-quadratic blind source separation Using NMF to unmix urban hyperspectral images. IEEE Trans. Signal Process. **62**(7), 1822–1833 (2014)

83. J.M. Mendel, Tutorial on higher-order statistics (spectra) in signal processing and system theory: theoretical results and some applications. Proc. IEEE **79**(3), 278–305 (1991)

84. F. Merrikh-Bayat, M. Babaie-Zadeh, C. Jutten, Linear-quadratic blind source separating structure for removing show-through in scanned documents. Int. J. Doc. Anal. Recognit **14**(4), 319–333 (2011)

85. A. Mohammad-Djafari, K.H. Knuth, Chapter 12. Bayesian approaches, in *Handbook of Blind Source Separation. Independent Component Analysis and Applications*, ed. by P. Comon, C. Jutten (Academic, Oxford, 2010), pp. 467–513

86. F. Mokhtari, M. Babaie-Zadeh, C. Jutten, Blind separation of bilinear mixtures using mutual information minimization, in *Proceedings of IEEE MLSP*, Grenoble (2009)

87. C.L. Nikias, J.M. Mendel, Signal processing with higher-order spectra. IEEE Signal Process. Mag. **10**, 10–37 (1993)

88. P. Paatero, U. Tapper, P. Aalto, M. Kulmala, Matrix factorization methods for analysing diffusion battery data. J. Aerosol Sci. **22**(1), S273–S276 (1991)

89. D.T. Pham, P. Garat, Blind separation of mixture of independent sources through a quasi-maximum likelihood approach. IEEE Trans. Signal Process. **45**(7), 1712–1725 (1997)

90. C. Revel, Y. Deville, V. Achard, X. Briottet, A linear-quadratic unsupervised hyperspectral unmixing method dealing with intra-class variability, in *Proceedings of the 8th Workshop on Hyperspectral Image and Signal Processing: Evolution in Remote Sensing (WHISPERS 2016)*, Los Angeles (2016)

91. J. Sigurdsson, M. O. Ulfarsson, J. R. Sveinsson, Blind nonlinear hyperspectral unmixing using an l_q regularizer, in *Proceedings of the 2018 IEEE International Geoscience and Remote Sensing Symposium (IGARSS 2018)*, Valencia (2018), pp. 4229–4232

92. A. Smilde, R. Bro, P. Geladi, *Multi-Way Analysis with Applications in the Chemical Sciences* (Wiley, Chichester, 2004)

93. Y. Su, J. Li, H. Qi, P. Gamba, A. Plaza, J. Plaza, Multi-task learning with low-rank matrix factorization for hyperspectral nonlinear unmixing, in *Proceedings of the 2019 IEEE International Geoscience and Remote Sensing Symposium (IGARSS 2019)*, Yokohama (2019), pp. 2127–2130

94. A. Taleb, A generic framework for blind source separation in structured nonlinear models. IEEE Trans. Signal Process. **50**(8), 1819–1830 (2002)

95. J.M.T. Thompson, H.B. Stewart, *Nonlinear Dynamics and Chaos* (Wiley, Chichester, 2002)

96. B. Yang, B. Wang, B. Hu, J.Q. Zhang, Nonlinear hyperspectral unmixing via modelling band dependent nonlinearity, in *Proceedings of the 2018 IEEE International Geoscience and Remote Sensing Symposium (IGARSS 2018)*, Valencia (2018), pp. 2701–2704

97. T.-J. Zeng, Q.-Y. Feng, X.-H. Yuan, H.-B. Ma, The multi-component signal model and learning algorithm of blind source separation, in *Proceedings of Progress in Electromagnetics Research Symposium*, Taipei (2013), pp. 565–569

Index

Printed in the United States
By Bookmasters